# THE TOLERANT SOCIETY

# THE TOLERANT SOCIETY

*Freedom of Speech and Extremist Speech in America*

Lee C. Bollinger



Oxford University Press · New York
Clarendon Press · Oxford
1986

Oxford University Press
Oxford New York Toronto
Delhi Bombay Calcutta Madras Karachi
Petaling Jaya Singapore Hong Kong Tokyo
Nairobi Dar es Salaam Cape Town
Melbourne Auckland

and associated companies in
Beirut Berlin Ibadan Nicosia



Published by Oxford University Press, Inc.,
200 Madison Avenue, New York, New York 10016



Library of Congress Cataloging-in-Publication Data
Bollinger, Lee C., 1946–
The tolerant society.
Includes index.
1. Freedom of speech—United States.
I. Title.
KF4772.B65 1986 342.73'0853 85–21410
ISBN 0–19–504000–7 347.302853



Printing (last digit): 9 8 7 6 5 4 3 2 1

Printed in the United States of America

*For Jean*

# Acknowledgments

No author should be more prepared to acknowledge his indebtedness to others than one who writes about the idea of free speech. It is a fundamental tenet of the field that dialogue is critical to the successful development of ideas—though, unfortunately, by no means a guarantee of success. All ideas are in some sense shared ideas; the benefits of criticism as a force for accuracy and clarity are not to be underestimated; and often we do not really know what we think until we have had the opportunity to say it to others—or, what is even more curious, until we have heard others say what we ourselves have tried to express.

In the course of working on this book, I have received help from many generous people, who performed the roles both of speaker and of listener, and it is a pleasure now to be able to thank them. To four individuals—Robert Burt, Ronald Gilson, Joseph Sax, and Frederick Schauer—all of whom read several drafts and listened to many, often inchoate, ideas, I am especially indebted. Several others read the final manuscript and gave me the benefit of their extensive comments: Francis Allen, Vincent Blasi, Yale Kamisar, Terrence Sandalow, Steven Shiffrin, Geoffrey Stone, and James B. White. There is no telling the debt I owe my wife, Jean, who did not have to read a word of the manuscript because she listened, at some point, to every line.

I have also received assistance from several institutions. I began this book during a year's leave in 1980, supported in part by a Humanities Fellowship from the Rockefeller Foundation. Three years later, in the collegial atmosphere of Clare Hall, Cambridge, I was able to complete a draft of the manuscript. That leave from my regular teaching responsibilities was supported by the William W. Cook Fund at the University of Michigan Law School.

The William and Mary Law School and its Institute of Bill of Rights Law invited me to deliver the George Wythe Lecture in November 1984, thereby making possible the first public airing of the theory I develop here. Much of the first two chapters will be found in a separate monograph published by the Institute of Bill of Rights Law under the title "Tolerance and the First Amendment." I should also like to indicate that a few sections of the first and third chapters have appeared in an earlier essay entitled "The Skokie Legacy: Reflections on an 'Easy Case' and Free Speech Theory," published in 80 Michigan Law Review 617 (1982). Portions of chapter five first appeared in "Free Speech and Intellectual Values," 92 Yale Law Journal 438 (1983), and are reprinted here by permission of The Yale Law Journal Company and Fred B. Rothman & Company.

*Ann Arbor, Michigan* L.B.
*September 1985*

# Contents

# THE TOLERANT SOCIETY

# Introduction

The origins of this book lie in a dissatisfaction with the current explanations and theories for the modern concept of freedom of speech, particularly as they apply to cases involving what we think of as extremist speech. This society has devised a remarkable legal principle under which highly subversive and socially harmful speech activity is protected against governmental regulation. People are at liberty to advocate the violent overthrow of the government, to urge the violation of legitimate and valid laws, to speak obscenities in public places, and to argue for discrimination against racial and religious groups. No other free society permits this kind of speech activity to nearly the same degree. Why should we in the United States? What do we as a society gain, or think we gain, by behaving in this fashion?

The historical development of this result is not so long-standing as many readers may think. Significant Supreme Court enforcement and interpretation of the constitutional principle of freedom of speech and press really dates only to the second decade of this century. Within that relatively short span of sixty or so years, the courts of this country—particularly the Supreme Court—have evolved an elaborate system of legal doctrines for the protection of speech activity. The jurisprudence of the First Amendment now consists of volumes of cases, hundreds of scholarly articles, and dozens of books. Law schools regularly provide not just one course but several courses on freedom of speech and press. The First Amendment landscape has both grown dramatically in size and been subdivided into many plots.

Traditionally, at the heart of thought about the First Amendment has been the matter of protection for extremist speech. Should it be protected and why? This question has formed the backbone of modern study of the First Amendment. In an age when the changes in telecommunications technology promise—or, from another point of view, threaten—to revolutionize our society, and raise in the process a plethora of free speech and press issues, the fundamental question of what to do with radical or extremist speech continues to provide the primary point of access for thinking about the First Amendment. Extremist speech has been the anvil on which our basic conception of free speech has been hammered out. It is the problem that first occupied the courts when they began applying the First Amendment, it continues to occupy their attention today and it promises to do so in the future. But, beyond its enduring presence as an issue for our social and legal institutions, it also has provided the primary context in which the general meaning of free speech has been defined.

This means, as a practical matter, that to think about the problem of how far the First Amendment should be extended is inevitably to end up thinking about the general theory of the First Amendment—about its basic purposes and functions for modern American society. To undertake the task of defining what appears, at least, to be the periphery is to gravitate toward considering the most fundamental issues about the principle itself. Inevitably, therefore, this gravitational field exerts its force on the argument of this book, which in the end proposes a new perspective from which to understand the contemporary meaning of the free speech principle.

Here again, on the matter of developing a theoretical structure for the First Amendment, the popular perception probably differs from the reality. Because the First Amendment to the Constitution, like the Constitution as a whole, has come to be viewed as primarily a *legal* document, to be interpreted by judges in the society, many people doubtless assume that as with all legal issues, an interpretive process is involved that differs radically from other methods of social decision making, such as the method employed in defining what level of welfare support should exist in the society. To some extent this is an accurate perception; however difficult it is to articulate the differences between "legislative" and "judicial" methods of analysis and decision making, it is widely understood that significant differences do exist. (In the course of this essay, we shall have occasion to consider at least some aspects of the nature of judicial, or legal, decision making.) It is also common to perceive legal reasoning as a far more certain and constricted enterprise than in fact it is—as any teacher of first-year law students knows. This is especially true of the First Amendment, that is, both the perception and the fact of the absence of certainty or preordained meaning.

In conventional legal analysis, both the text and the original intention of the authors of the text are treated as primary sources of meaning. Despite the attempts of many judges and lawyers to make these sources appear to be clear guides, it is a fact of First Amendment life that they have had little to do with the actual development of that principle. As for the original design of the Founding Fathers, there is precious little evidence to reveal their intentions, and what evidence there is suggests something far different from, and more limited than, what has evolved in this country in this century. Besides, with a *constitution*, there is good reason to feel less bound by original understandings because of the desirability of having a basic charter that is capable of adapting to the needs of each age—which, of course, may well have been precisely the original understanding. As for the specific language of the First Amendment—"Congress shall make no law . . . abridging the freedom of speech, or of the press"—it is far less "absolute" a prohibition than it might at first appear. The language has a deceptive clarity. While it does say "no law," it also says "freedom of speech," and since no one has ever been willing to contend that every speech act under all circumstances should be protected, we must necessarily embark on a decisional process of exclusion and inclusion for which we very quickly find we need some theoretical guidance beyond what the language itself offers.

The truth is that judges and legal writers have been left largely adrift on a sea of possible interpretations of the First Amendment, and they understandably have often been inclined to clutch at anything within reach with the least appearance of logical buoyancy. This is not to say, however, that the process of defining free speech has been without intellectual foundation or integrity, only that it has not been predetermined or fixed. The process that has occurred resembles the development of legal principles in the traditional common law system. Different courts and judges take stabs at reasoning their way through the tangles of specific disputes, which over time become an aggregate body of experience from which comparisons can be drawn and assessments made. In such a process, experience and time are the great validators and the givers of higher meaning. Through a largely indecipherable process, various fragments of statements in judicial opinions and other writings, and the factual narratives around which those statements were made, are plucked from the ever-increasing mass of material; these take on the status of a higher text, of near biblical significance, to which nearly everyone regularly refers in trying to provide a coherent and intelligible account of the field.

Such a process may, however, complicate our efforts to develop an understanding of the social functions performed by the free speech idea. There can be a kind of artificiality, an excessive neatness of meaning, that this winnowing process gives to the flux of events, producing the kind of dissatisfaction we often feel toward program notes that try to explain a symphony. Moreover, the rhetoric itself may remain the same while the social functions change, thus serving to interfere with or divert understanding. While classicists tell us that the free speech principle at best played only a very limited role in the thought of ancient Greece and Rome, the life history of the principle dates back several centuries, at least to the seventeenth century, which is when John Milton wrote his famous defense of liberty of speech and press in *Areopagitica*. The contemporary rhetoric of free speech, the language and terms used to think and talk about the principle, draws surprisingly heavily on the writings of these earlier centuries. Yet, it is entirely possible, even probable, that the social and political functions of the free speech concept have changed significantly over the centuries. Just as the concept of wilderness had a very different personal and social significance for the early settlers of this country than it does in today's highly urbanized and organized society, so the free speech principle may well have acquired new meanings. There is a serious risk, therefore, that in trying to understand the contemporary significance of free speech, we will be misled into assuming a continuity of real meaning from the continuity of language used to talk about free speech.

This points to yet another obstacle in the path of achieving an understanding of the roles played by the concept of free speech in contemporary American society. While the interpretive authority over the free speech principle has become part of the legal realm, the free speech idea nonetheless remains one of our foremost *cultural* symbols. It is suffused with symbolic significance. As for its practical impact on our own speech behavior as individuals, Mark Twain's notorious comment on the subject retains its sting: "It is by the goodness of God," he said, "that in our country we have those three unspeakably precious things: freedom of speech, freedom of conscience, and the prudence never to practice either of them" (Pudd'nhead Wilson's New Calendar, chap. 2).

Most of us are probably an easy mark when it comes to assessing the degree to which we take advantage of the opportunity to express our opinions, or for that matter even to exercise independent thought. But, at the same time, it is also the case that some members of the society may be said to abuse the available liberty by saying publicly things that violate the most basic canons of human decency. Yet, oddly enough, it seems to be partly because such speech is protected that the free speech idea holds such a peculiar and powerful fascination for us. In any case, the idea constitutes an important piece of the American character, for which we have countless and almost daily pieces of evidence. And we ought to expect this broader social symbolism to interact in important ways with the strictly "legal" process of interpretation applied to the principle, though to define that interaction will doubtless be as elusive a task as it is important to undertake. Each—the legal analysis and the cultural symbolism—may be expected to influence the other in a reciprocal relationship.

This, at least, is a guiding hypothesis of this essay about the modern significance of the free speech principle. The thesis offered here is that free speech has, indeed, developed new meaning over the course of the last several decades, a new significance that helps especially to account for the extremes to which the principle has been taken. For a variety of reasons, however, that newer meaning has not emerged clearly in the thinking, or at least the discussion, about the principle; and the course of the argument in the chapters that follow is to show how that new function flows naturally out of traditional patterns of theoretical thought about the free speech principle.

Traditionally, the focus of free speech theory has been on identifying the ways in which having the freedom of speech is valuable to us; just as it would be natural to point to the paintings of the great masters to explain why we should permit the activity of painting, so we have been inclined to point to the uses of speech we most value to justify the free speech principle. Attaining truth, exercising our democratic prerogative of self-government, and satisfying our yearnings for self-expression are interests that nearly everyone regards as vital and that can be readily, and reasonably, associated with the activity of speaking. Free speech theory has, in short, been highly successful in uncovering the riches of a policy of treating speech as a zone of liberty.

Yet, a disturbing lacuna appears in our theory as we confront the reality of protecting extremist speech (or what might be referred to as the overprotection of speech). For many of us who regard ourselves as committed to the idea of freedom of speech, the benefits traditionally associated with such a principle—primarily related to the process of discovering "truth"—begin to break down as the speech in question strikes more and more deeply at the personal and social values we cherish and hold fundamental to the society. While other considerations are often brought to bear on the free speech side in these cases (considerations we shall take up later under the heading of the "fortress model"), they too raise distressing implications. Despite the inability of conventional theories of free speech to account for this phenomenon of overprotection, there does seem to be a shared intuition that the society adds something important to its identity, that it is significantly strengthened, by these acts of extraordinary tolerance. It is the purpose of this book to try to explore the elements of that intuition; in that inquiry we will be able to reformulate the aims of the modern-day free speech idea.

It may be said of the new perspective offered here that while free speech theory has traditionally focused on the value of the activity protected (speech), it seeks a justification by looking at the disvalue of the response to that activity, by directing our attention to the problematic character of the feelings evoked by, in particular, extremist speech. Such troublesome feelings, however, are not to be understood as the affliction of only a segment of the society but rather as being universal. Nor, on the other hand, can it be said that they always govern our responses to this kind of speech activity. One of the significant points of departure between the perspective taken here and at least some prior theoretical accounts of free speech is the willingness to take as a working premise the idea that a good part of the speech behavior we are talking about is often unworthy of protection in itself and might very well be legally prohibited for entirely proper reasons. To acknowledge that, however, does not mean that a choice to tolerate such speech is irrational or unwise. The rationality and wisdom of choosing the course of tolerance can be derived from a neglected insight—namely, that the problematic feelings evoked by this kind of speech activity are precisely the same kinds of feelings evoked by a myriad of interactions in the society, not the least of which are the reactions we take toward nonspeech behavior. Thus, while previous theoretical accounts of free speech have been led to focus on the differences between speech and other behavior (often described as "conduct"), the theory described here locates important social meaning in the similarities between the responses generated by all kinds of behavior. At this stage in our social history, then, free speech involves a special act of carving out one area of social interaction for extraordinary self-restraint, the purpose of which is to develop and demonstrate a social capacity to control feelings evoked by a host of social encounters.

I have chosen the term *tolerance* to describe this capacity sought, as well as the term *intolerance* to describe an incapacity; a few words about them are necessary because of their imprecision for these assignments. "Tolerance," of course, is often used in an entirely neutral sense: simply the ability to endure. While I do not intend to use the term so broadly, its breadth is importantly suggestive of the shift in perspective I will be offering. To the extent that the capacity I speak about here has far wider relevance to social interaction than to the narrow pursuits traditionally thought to be advanced by the free speech principle, that sense of enlargement is attractively conveyed by the word. Furthermore, insofar as tolerance is often associated with the ability to coexist with different beliefs, that too is an advantage because it is the phenomenon of belief and of our attitudes toward belief that lies at the core of the troublesome feelings we find at issue in speech cases. But the most favorable attribute of all of the term is that we regularly use it to describe our reactions to nonspeech as well as speech behavior. (Tolerant means "showing understanding or leniency for *conduct* or ideas . . . conflicting with one's own" [my emphasis], in the words of *Webster's Third New International Dictionary*.) If anything is central to the thesis developed here, it is that the commonality or relationship of the two areas of behavior should interest us at least as much as their divergence.

It is important, however, that we not be misled by the use of the term *tolerance*. In a sense, everyone who thinks about free speech also thinks about, and tries to account for, why speech activity should be "tolerated." Indeed, everyone who has already proposed a theory of free speech might properly have referred to it as the tolerance theory. But, putting aside the question of proprietary rights, the label applied should not divert attention from the differences in perspective offered by competing or complementary theories.

Nor would I want the reader to assume that the view taken here is premised on the notion that tolerance is always a virtue and intolerance a vice. There are times when tolerance constitutes moral weakness and is itself properly to be condemned, just as there are times when responding "intolerantly" is a sign of admirable moral strength—though because we usually use the term *intolerance* pejoratively, we express ourselves on these occasions negatively, saying we must "not tolerate" something when that is regarded as the appropriate response. I try at various points in the text to discuss the complexity of the personal issues both suggestively raised and sometimes obscured by these terms, but I hope the reader will keep the ambiguity of the terminology in mind at all times.

One final point: Although the analysis here includes discussions of many free speech cases and doctrines, the primary focus is on how we think or ought to think about free speech; a theoretical perspective is offered from which to approach the tasks of devising doctrines and deciding cases. In that sense the inquiry is preliminary, though still essential. Future debate on the merits of the free speech principle we now have must begin with an understanding of the goals of free speech. The purpose of this book is simply to contribute to that understanding.

# 1

# Enslaved to Freedom?

There is a curious disjunction in our attitudes about the degree to which we should tolerate the speech of others. When we compare our reluctance to impose *legal* restraints against speech with our readiness to employ a host of informal, or *nonlegal*, forms of coercion against speech behavior, the paradox is striking. If a person expresses some view we find deeply offensive—say, for example, by making a racially derogatory comment—we will probably insist on censure of some kind and feel guilty if none occurs. To implement our unofficial decree, we may draw on a myriad of coercive responses typically at our disposal: We may respond with ridicule or humiliation; we may practice any number of forms of social shunning; or we may withhold various practical benefits, like employment opportunities. When the offender is a person who holds public office, we will insist that resignation follow. To be told that we ought to restrain ourselves from ever, in any way, coercing or penalizing any person for what that person says would strike us not only as bizarre but as plainly wrong. We do what we can to create an environment in which such expressions are deterred and transgressions are visited with appropriate penalties.

Within the special realm of the constitutional right of free speech, on the other hand, our response is quite different. As soon as someone proposes making it unlawful to say something offensive, and to assess criminal or civil penalties for any violations, the general free speech principle will be invoked, and in all likelihood the whole plan will be tossed aside as unacceptable. More to the point, the degree to which we exercise restraint in the use of legal penalties against speech activity is not just marginally different from our employment of nonlegal forms of coercion; on the contrary, the difference in approach in each area is substantial and in some cases dramatic.

This was revealed for us with painful clarity just a few years ago in a widely publicized case involving a Nazi group that claimed the "right" to conduct a march in a suburb of Chicago called Skokie. At the time, the population of Skokie included some forty thousand Jews, several thousand of whom were direct survivors of the World War II concentration camps. This, of course, was not lost on the self-proclaimed Nazis. Nor was their purpose lost on those who stood witness to the projected event. It was perfectly clear to everyone that a primary aim of the Nazis was to inflict as much insult and fear as they could on the Skokie residents, and no doubt on other Jews as well. And yet, the results reached by the two highest courts that considered the issues—a federal court of appeals and the Illinois Supreme Court—were the same: this "speech activity" was protected by the First Amendment to the Constitution.[1]

Why is it, therefore, that one form of coercion or punishment—legal restraints—has essentially been removed from our general armory of possible responses to speech we hate and fear and believe dangerous to the values we cherish? Is it because we think the effects of legal restraints are disproportionate to the harm caused? Or that the risks of vesting coercive power in the government are too great?

These are complicated questions, and in succeeding chapters we shall consider them in some detail. But it is important at the outset to see that the answer to the disjunction is not so self-evident as many of us who think of ourselves as committed to the First Amendment may implicitly assume. Many speech acts cause more harm than does much nonspeech conduct that we regulate with the sanctions of fines and imprisonment, and we view it as necessary with respect to nonspeech behavior to entrust the government with the authority to maintain a proper degree of social order. Indeed, we commonly speak of the "duty" of the state to perform this task.

But perhaps the strangest aspect to the paradox, or disjunction, I have noted lies not in the fact that we seem to treat legal and nonlegal punishments differently but rather in the fact that once any issue is cast in First Amendment form, our attitude toward the pros and cons of using coercion seems so markedly different. Under the principle of free speech, we celebrate self-restraint, we create a social ethic of tolerance, and we pursue it to an extreme degree. Our attitudes seem not to correlate with any sober assessment of the possible incremental gain in the protection of our own freedom. It is therefore in the extremeness of our enthusiasm for what is done in the name of the free speech principle that the sense of disjunction primarily arises.

At the same time, for many people it is, and has been, this extremeness that is the most inexplicable and troublesome feature of free speech in the United States. They can readily understand the sense of limiting the use of governmental power to regulate general discussion within the society. They also have the sense that there must be limits to any principle and that somehow free speech has been taken far beyond those limits. The disjunction in our attitudes toward legal and nonlegal coercion against expression, otherwise mostly unnoticed, is so heightened in the extreme cases as to make it obvious to everyone. That occurred in the *Skokie* case.

Few legal disputes in the last decades caught the public eye with such dramatic power as did that case. For well over a year, as the case moved ponderously through the courts, it was seldom out of the news and often on the front pages of newspapers when it was in the news. When the American Civil Liberties Union (ACLU) took up the legal defense of the Nazis, its membership rolls gave telling evidence of the public dissatisfaction, even incredulity, at the free speech position in the case. Thirty thousand members resigned their membership, at an annual cost in lost revenues to the organization of half a million dollars.[2] To many people this was not freedom of speech, it was the abuse of a liberty, the license to inflict harm on other people. Even if one viewed the Nazi aim in more modest terms, as that of only establishing a fascist regime, the assertion of a free speech right seemed only to raise a profound paradox: Why, after all, should a free speech principle be extended to those who would use it to advocate the destruction of that liberty? That this was an old puzzle made it none the less compelling. Here individuals were advocating an ideology the country had invested incalculable resources only a few decades ago to defeat; now it was being protected in its efforts to resurrect itself. Surely, many wondered, there is something disturbingly anomalous about that.

## I

Actually, these reservations about the extension of the free speech principle to cover this kind of extremist speech were but echoes of a similar indictment heard at the very inception of the modern free speech principle, and it is useful to return to that indictment as a means both of clarifying the reservations and of obtaining a sense of the continuity of this primary concern about the nature of the principle. The attack was by John Wigmore, a law professor and dean at Northwestern University Law School and a scholarly figure of major stature. The object of Wigmore's indictment was Justice Oliver Wendell Holmes's dissent in *Abrams v. United States*[3]—a dissent that was to provide the underpinnings of the contemporary free speech principle, as later applied to controversies like *Skokie*.

*Abrams* involved the prosecution of five Russian aliens for distributing leaflets in New York City in August 1918. These leaflets praised the Russian Revolution, denounced President Woodrow Wilson for attempting to intervene and reverse the successes of the Communists, and urged the workers in the United States (particularly munitions workers) to protest by engaging in a general strike. The Russians were prosecuted under the Espionage Act,[4] a World War I piece of legislative handiwork that proscribed a variety of activities the Congress had deemed potentially harmful to the war effort.

A majority of the Supreme Court upheld the convictions, finding no violation of the First Amendment. For precedent, these justices relied heavily on three cases decided earlier in the same year, which ironically had been authored on behalf of the Court by Holmes himself. The first of that trilogy (and the first important Supreme Court decision on the First Amendment since its adoption), *Schenck v. United States*,[5] was in many important respects seemingly very similar to *Abrams*. Schenck had been the general secretary of the Socialist party, and he, along with another member of the party's executive board, was charged with having distributed some fifteen thousand leaflets in which it was argued, in "impassioned language," that the conscription law was immoral and unconstitutional and that people should resist. "In form at least," Holmes wrote, "[it] confined itself to peaceful measures."[6]

This prosecution was also under the Espionage Act. Holmes dealt with the case by pointing out the necessity of drawing some limits on the free speech principle. While, he said, the Court was prepared to "admit that in many places and in ordinary times the defendants in saying all that was said in the circular would have been within their constitutional rights,"[7] the actual scope of the protection afforded by the First Amendment depended on the exact context in which the speech occurred. Thus, in words now immortalized: "[T]he most stringent protection of free speech would not protect a man in falsely shouting fire in a theatre and causing a panic. It does not even protect a man from an injunction against uttering words that have all the effect of force."[8] The guiding principle for Holmes, therefore, was "whether the words used are used in such circumstances and are of such a nature as to create a clear and present danger that they will bring about the substantive evils that Congress has a right to prevent."[9] To Holmes "[i]t was a question of proximity and degree."[10]

Applying this principle to the *Schenck* prosecution, Holmes appeared to have little difficulty in finding for the government. "When a nation is at war," he cautioned, "many things that might be said in time of peace are such a hindrance to its effort that their utterance will not be endured so long as men fight and that no Court could regard them as protected by any constitutional right."[11]

The majority in *Abrams* seemed to think similarly. Holmes, however, now did not, and the reason for his seeming turnabout has been a matter of controversy and speculation ever since.

In his *Abrams* dissent, Holmes began by reaffirming his votes in the earlier trilogy of cases of which *Schenck* was one, his commitment to the "clear and present danger" test he had developed in *Schenck*, and, finally, his earlier stated belief that congressional power to limit speech "undoubtedly is greater in time of war than in time of peace because war opens dangers that do not exist at other times."[12] But now, instead of starting from the negative pole of what free speech did *not* protect and working backward to the speech at hand, he began from the positive side and worked forward. Even in wartime, Holmes argued, "Congress certainly cannot forbid all effort to change the mind of the country."[13] Holmes was quick to make clear, however, that he shared no sympathy with "these poor and puny anonymities," whose beliefs he labeled a "creed of ignorance and immaturity."[14] Given this puniness, the sentences of twenty years' imprisonment seemed to Holmes particularly disproportionate; only the "most nominal punishment,"[15] he said, was called for under the circumstances.

But Holmes's enthusiasm for recognizing a "right" at stake in this case (rather than simply finding an appropriate scale of punishment) seems to gain force even as he writes. By a few phrases the mood of the opinion shifts dramatically and Holmes now sets forth, in memorable words, his primary argument for free speech:

> But when men have realized that time has upset many fighting faiths, they may come to believe even more than they believe the very foundations of their own conduct that the ultimate good desired is better reached by free trade in ideas—that the best test of truth is the power of the thought to get itself accepted in the competition of the market, and that truth is the only ground upon which their wishes safely can be carried out. That at any rate is the theory of our Constitution.[16]

In the concluding sentences, Holmes lent eloquence to his words by regretting his lack of eloquence, closing with an apology: "I regret that I cannot put into more impressive words my belief that in their conviction upon this indictment the defendants were deprived of their rights under the Constitution of the United States."[17]

These were the sentiments, and not those of Holmes in the *Schenck* case, that proved to be the auspicious beginning for the principle of free speech as we think of it today. Though spoken in dissent, eventually they were to take root and find their logical fulfillment in later cases, of which *Skokie* is the most prominent in our time. To be sure, they have not been consistently adhered to; many regard the McCarthy era cases as constituting a significant abandonment.[18] Nevertheless, within the legal community today, the *Abrams* dissent of Holmes stands as one of the central organizing pronouncements for our contemporary vision of free speech. And the scope of the shelter it extends to speech activity is very wide indeed, for under it the First Amendment protects against legal interference all speech activity until the point at which it "so imminently threaten[s] immediate interference with the lawful and pressing purposes of the law that an immediate check is required to save the country."[19] Until that grave crisis point has been reached, no legal action may be taken.

To this definition of the scope of free speech, Wigmore took strenuous objection in an article published in the Illinois Law Review shortly after the *Abrams* decision. The essay is written in a strident tone, reflecting all the heat of a still-flushed veteran returning from World War I, in which Wigmore had served as a colonel. Wigmore's hostility to what Holmes had done is undisguised and is most clearly represented by his refusal in the course of the article to refer to Holmes by name; he speaks only of "the dissent." The article suffers from its unmeasured critical tone. But there is an emotional power to the essay, which makes it one of the most extraordinary documents we have in the free speech literature. This power derives from Wigmore's outrage at Holmes's own emotionally stirring rhetoric. In the end, Wigmore presents a challenge that must be met squarely if we are ever to have a viable theory of freedom of speech.

Much of Wigmore's concern in the article, as one would expect, is with the specific facts involved in the *Abrams* case. Holmes, he declares, had grossly minimized the risk to the country from the defendants' speech. By August 1918, he says, the United States had only begun its efforts to shore up the nearly defeated Allied countries, and the success or failure of that intervention was still a matter of considerable doubt. The outcome depended greatly on the still-uncertain capability of the country to produce the munitions needed to supply the soldiers then in the trenches in France—which, of course, it had been the aim of the *Abrams* defendants to disrupt. "The national agonies" of these months, Wigmore charged, "were apparently unsensed by the Minority Opinion."[20] Holmes had been "blind to the crisis—blind to the lasting needs of the fighter in the field, blind to the straining toil of the workers at home, obtuse to the fearful situation which then obsessed the whole mind and heart of the country."[21]

Wigmore's article becomes more interesting, however, as it moves from the particular to the general. In Wigmore's view, Holmes's failing was not just a momentary lapse of judgment; it was the manifestation of a generally misguided and distorted social vision posturing under the banner of freedom of speech. The real significance of Holmes's dissent, therefore, was in its representation of a more widely held state of mind. Many people, Wigmore asserted, were tending to overemphasize the value of the liberty, and the need to secure it against the threat of government censorship, and correspondingly to underemphasize the costs and dangers that unrestricted speech would bring in its wake. "And so the danger now is," Wigmore wrote, "rather that this misplaced reverence for freedom of speech should lead us to minimize or ignore other fundamentals which in today's conditions are far more in need of reverence and protection. Let us show some sense of proportion in weighing the several fundamentals."[22]

Wigmore saw this tendency to underestimate the true social harm from speech behavior not just in Holmes's apparent disregard for the needs of the war effort but also in the way Holmes chose to characterize the nature of the defendants' acts. In the course of his dissent, Holmes had argued that it was pointless to be so concerned about "the surreptitious publishing of a silly leaflet by an unknown man,"[23] that these defendants were but "puny anonymities," and that it was unlikely anyone would have been persuaded by the defendants' urgings. Wigmore, conceding that the defendants' acts by themselves were unlikely to harm the war effort, took Holmes to task for thinking about and analyzing the risks associated with the defendants' acts in a highly idiosyncratic and peculiar way. Ordinarily, Wigmore pointed out, the society is not denied the power to punish those who set out to commit crimes, to engage in behavior that the legislature has deemed as a general matter to be socially harmful, but for one reason or another either fail in the effort or accomplish the deed without the adverse social consequences. It is no defense to say that you were ineffectual in your attempt to commit a wrong: "The relative amount of harm that one criminal act can effect is no measure of its criminality, and no measure of the danger of the criminality."[24] We may legitimately wish to avoid making case-by-case assessments of actual injury; in any event, one person's attempt may prompt others to act, and a large number of acts, which are insignificant in themselves, may together produce a very large harmful effect. We comprehend that reality and, on that understanding, structure our system of criminal law; why, then, should we not do so here as well, Wigmore asked.

For Wigmore the other side of the coin was the tendency to overemphasize the need to secure the liberty of speech against restriction. Again, as with the tendency to minimize the risks of speech behavior, such thinking was not aberrational but unfortunately commonplace. Wigmore saw as anachronistic and foolish, like "Don Quixote, fighting giants and ogres who have long since been laid in the dust,"[25] the quest to erect such an elaborate shelter for free speech:

> After all, is not this tenderness for the right of freedom of speech an over-anxiety? Is not this sensitive dread of its infringement an anachronism? Has not the struggle for the establishment of that freedom been won, and won permanently, a century ago? Do we not really possess, in the fullest permanent safety, a freedom and a license for the *discussion* of the pros and cons of every subject under the sun? Simply as a matter of "free trade in ideas," is there not in Anglo-America today an irrevocably established free trade in every blasphemous, scurrilous, shocking, iconoclastic, or lunatic idea that any fanatical or unbalanced brain can conceive? And is there any axiom of law, constitution, morals, religion, or decency which you and I cannot today publicly dispute with legal immunity?[26]

This oversensitivity to the dangers to free speech meant that the right was "being invoked more and more in misuse," for and on behalf of "impatient and fanatical minorities—fanatically committed to some new revolutionary belief, and impatient of the usual process of rationally converting the majority."[27]

As Wigmore then set about defining the proper limits of the free speech principle, he sought to minimize the disjunction to which I referred at the outset between our personal and our constitutional thinking about those limits, trying to bring the latter more in line with the former. In the "abnormal" situation of wartime, which was true of the *Abrams* case, Wigmore found the proper line by defining what was an appropriate "moral" response to the *Abrams* type of expression. He found that the "moral right of the majority to enter upon the war imports the moral right to secure success by suppressing public agitation against the completion of the struggle."[28] The minority must respect the decision of the majority, which has been reached by their following proper procedures. So, the state, just like "any well-meaning citizen," has a "moral right . . . to smite . . . on the mouth" any "disaffected citizen" who was to go among a "company of soldiers in war-time on their way to the front . . . [and] harangue: 'Boys! this is a bad war! We ought not to be in it! And you ought not to be in it!' "[29] To believe otherwise, in Wigmore's view, is to favor freedom at the expense of a nation's inherent right to govern itself and to ensure its own survival.

Wigmore's charges against Holmes's position on the scope of the First Amendment have been repeated in subsequent decades, whenever First Amendment protection for extremist speech has been sought; seldom, though, have the challenges been expressed with as much force as Wigmore was able to give them. Is it not anomalous to take a position that concedes it is legitimate for the society to restrain harmful *conduct*, yet insists at the same time that the society should and can do nothing about the advocacy of such harmful behavior until just before the moment when the advocacy is about to turn into action—essentially the position on the "clear and present danger" test? Is it not equally anomalous and paradoxical to hold that we should extend First Amendment protection to those who wish to use that right to destroy the First Amendment itself, or the basic democratic institutions that provide the necessary foundation for that right as well as other rights? If we believe in anything in this society, our beliefs must include the First Amendment, so why should the society be compelled to tolerate speech acts committed to its destruction?

These asserted anomalies and paradoxes, however forceful in their logic, are only derivative of a deeper challenge to conventional free speech thinking that we find in the Wigmore essay: that we are in danger of losing our sense of proportion, of balance, in our thinking about free speech. It is this question about a loss of a sense of reasonableness, of succumbing to a fascination with purity and simplicity and a wish for an unambiguous universe, that is most troublesome.

Free speech has become such a fixture of the American identity that our critical faculties may be unconsciously suspended when we are in its presence, an ironic result given the commonly understood purpose of the principle to remove the shackles on dissent and to encourage openness of mind. Perhaps, however, this charge of having slavishly pursued a classical idea to the point where it has lost its original meaning and human value, to where imagination and creativity are deadened by strict extension, makes free speech but a part of other features of twentieth-century life. In a sense, Wigmore's comments are typical of those one often hears about the developments in modern art, or perhaps, to take another field, particle physics. Whatever the nature of the broader pattern, it has been charged that free speech has come to be a principle without proper proportion, and the strength of the charge is attested to by its repetition as each new *Abrams*-wrought case has arisen and has been decided according to the logic of Holmes's dissent.

Have we, it must therefore be asked, fallen victim to what the intellectual historian Isaiah Berlin called the "suffocating straitjackets" of "great liberating ideas"?

> The history of thought and culture is, as Hegel showed with great brilliance, a changing pattern of great liberating ideas which inevitably turn into suffocating straitjackets, and so stimulate their own destruction by new, emancipating, and at the same time, enslaving conceptions.[30]

## II

This was certainly the uneasy sense, however inarticulate, many people had about the free speech position in the *Skokie* case—the sense of loss of judgment and of the ability to draw reasonable lines, to assess fairly the risks and costs of speech and the risks and costs to free speech of imposing limits. As it had for Wigmore in 1920, the disjunction became too great; the free speech position appeared unjustified, and it raised disturbing implications.

Just what did the courts that considered the *Skokie* case, and that reached a favorable result for free speech protection, offer by way of defense and explanation of that position? We should examine the opinions to see if in the intervening years our free speech theory has generated a justification for the extremes to which it has quite obviously been taken, or a response to Wigmore's charges.

To understand the judicial positions taken in the *Skokie* dispute, however, we must pause a moment to recount the salient facts of the case. The details need not detain us long, but a sketch will be helpful, particularly since in subsequent chapters we shall return to the case as illustrating other points.

The controversy began in the spring of 1977 when Frank Collin, the leader of the Chicago-based National Socialist Party of America (NSPA), requested a permit to march in front of the Skokie village hall. The proposed march was to be held in the midafternoon and to take about half an hour, during which time approximately three dozen members would march in single file in front of the village hall. The group, according to Collin, would be wearing Nazi-style uniforms, which he described in chilling detail: "We wear brown shirts with a dark brown tie, a swastika pin on the tie, a leather shoulder strap, a black belt with buckle, dark brown trousers, black engineer boots, and either a steel helmet or a cloth cap, depending on the situation, plus a swastika armband on the left arm and an American flag patch on the right arm."[31] Collin's request came in March 1977.

After receiving this notification of the intended demonstration, Skokie filed suit in the local circuit court, seeking an injunction against it. As this litigation worked its way up and down the line of state courts, another suit was reaching the federal courts. On May 2, 1977, the march having been temporarily forestalled by the state court suit, the city enacted three ordinances purporting to cover all marches and demonstrations. The first provided various requirements for the issuance of any permit for parades and marches in excess of fifty persons. Insurance had to be obtained ($300,000 in public liability and $50,000 in property damage) and assurances given that the group "will not portray criminality, depravity or lack of virtue in, or incite violence, hatred, abuse or hostility toward a person or group of persons by reason of reference to religious, racial, ethnic, national or regional affiliation." The second and third ordinances were specifically criminal laws, violation of which could be punished by a fine of up to $500 or imprisonment of up to six months. One forbade the "dissemination of any materials within the Village of Skokie which promotes and incites hatred against persons by reason of their race, national origin, or religion, and is intended to do so." "Dissemination of any materials" was defined to include "display . . . of signs" and "public display of markings and clothing of symbolic significance." The final ordinance prohibited the wearing of "military-style" uniforms during any public demonstration.[32]

Following the enactment of these ordinances, Collin applied for a permit to engage in a march similar to that proposed earlier but now rescheduled for the Fourth of July. The Skokie authorities denied the permit, indicating that the march would violate the last of the three ordinances. At this point the American Civil Liberties Union filed suit on behalf of the NSPA in the federal district court, contending that all the ordinances were unconstitutional under the First Amendment and seeking declaratory relief to that effect. By the time the case reached the Court of Appeals for the Seventh Circuit, however, Skokie had conceded that the insurance requirement (at least as applied to the proposed march here) and the anti-uniform prohibition were unconstitutional. At issue in that court, therefore, was the city's attempt to prohibit the promotion or incitement of hatred "against persons by reason of their race, national origin, or religion." The Illinois Supreme Court, on the other hand, which was the ultimate state court to rule on the lawsuit originally filed in the state circuit court, was presented with a general claim by the city of Skokie that it had a right to stop the proposed march without regard to any particular ordinance provision. Following decisions by both courts, both adverse to the city of Skokie, a petition for review was filed in the United States Supreme Court. That Court, however, declined to take the case for decision (two justices, Blackmun and White, dissented from the refusal).[33]

Now, we are interested in knowing how these courts justified the conclusion, arrived at independently in each case, that the community of Skokie was precluded by the First Amendment to the Constitution from stopping a group of Nazis from demonstrating in front of its village hall. If, in answer to Wigmore, this was not the distorted interpretation of a basically sound, classical idea, then why not? Have the intervening decades brought forth an answer to that still relevant question? It is to the opinions of the courts that we must look for answers.

One of the striking characteristics about the *Skokie* case is that one encounters confusion and uncertainty wherever one turns. To look at *Skokie* is to enter a hall of mirrors, where image and form, appearance and reality, are continually confused. Take first the parties. The "Village" of Skokie is not a village in any meaningful sense of the term, though it was obviously to its advantage to portray itself as such to enhance our sense of the intrusion into the community by the proposed march. It is, simply, a Chicago suburb. Even the question of whose "turf" it is, is a matter of some uncertainty. Before World War II Skokie had primarily been a German community, known as Little Germany, and the home of the German-American bund.[34] On the other side, the National Socialist party, with its few dozen members, was hardly a "party" at all, though it too no doubt regarded the self-depiction as advantageous for its public relations. Nor was the extent of its identification with the policies of the Third Reich entirely clear. Even the real identity of its leader, Frank Collin, was a matter of doubt. Symbolic of the deeply confusing nature of the dispute, it appeared that Collin's father was a Jew and a survivor of Dachau.[35]

This problem of fixing one's vision on the true reality extended to the legal issues. Was this a march to proclaim religious and racial hatred, or even genocide? No, said the Nazis quite explicitly from the beginning. It was to protest the denial of their "free speech rights." The placards they proposed to carry were to be inscribed with the words White Free Speech and Free Speech for White America, a protest against the demands for an insurance policy as a prerequisite to obtaining a march permit. On the other hand, it was possible, and not implausible, to read a negative implication into the slogans, to the effect that only whites, and then only *some* whites, should be accorded First Amendment rights. Furthermore, the placards were not the only communicating objects that would be present; there was also the storm trooper regalia Collin's group would be wearing with its own deadly messages. Finally, beneath this dispute about what the group intended to say was yet another source of confusion, namely, whether the asserted desire to march was itself fictitious. At times it seemed that the object was not to march but to be opposed in the effort. In fact, the Nazis never did march in Skokie, even after they had secured the right to do so; instead, they chose to make a brief appearance at the plaza of the Daley Center in downtown Chicago, where before the Skokie dispute had even arisen they had been similarly rebuffed in their efforts to obtain a march permit.[36]

This problem of determining who these people were and what they were fighting about, of separating appearance and reality, is endemic to the case as a whole. Not only does it involve issues of the true identities of the parties and the central dispute between them, including their actual intentions and feelings about the conflict, but the legal positions formulated by each side as well. Getting to the bottom of things is a more arduous task than one might at first suppose. This is certainly true when one attempts to decipher the judicial opinions in order to discover their motivating rationale.

On one issue, at least, the opinions are unmistakably clear, namely, that the judges wished it known that they personally repudiated the ideas held by Collin and his group. Virtually every opinion written in the case contains somewhere within it such a personal statement by the judges. These denouncements are unqualified. They unambiguously proclaim the Nazi ideology a collection of monstrous errors. The opinion of the federal court of appeals begins with the following statement: "We would hopefully surprise no one by confessing our personal views that NSPA's beliefs and goals are repugnant to the values held generally by residents of this country, and, indeed, to much of what we cherish in civilization."[37] It next refers to a recent trial involving the revocation of citizenship of an alleged Nazi war criminal and the television production of a movie about the Holocaust, and adds: "We cannot then be unmindful of the horrors associated with the Nazi regime of the Third Reich, with which to some real and apparently intentional degree appellees associate themselves."[38] Finally, at the very end of the opinion the judges return to the same theme, with an even stronger denunciation:

> The preparation and issuance of this opinion has not been an easy task, or one which we have relished. Recognizing the implication that often seems to follow over-protestation, we nevertheless feel compelled once again to express our repugnance at the doctrines which the appellees desire to profess publicly. Indeed, it is a source of extreme regret that after several thousand years of attempting to strengthen the often thin coating of civilization with which humankind has attempted to hide brutal animal-like instincts, there will still be those who will resort to hatred and vilification of fellow human beings because of their racial background or religious beliefs, or for that matter, because of any reason at all.[39]

In this way, then, the designated "legal" analysis of the case is bracketed by these clear, uncomplicated, "personal" resolutions of the same issues.

There is more to these personal proclamations, however, than first meets the eye. To most of us these statements would no doubt appear perfectly unremarkable; indeed, their absence would have been regarded as cause for concern, since an indication of personal neutrality on the ideas of Nazism, which might arise from that silence, would have presented a shockingly unexpected state of affairs. But that is only because we agree with the conclusion that Nazism is horrendously evil. Believing so, we are less sensitive to seeing how coercive and threatening such judicial denigrations of the speech protected can be. What we see, instead, is the satisfaction of the individual needs of the judges to dissociate themselves from the beliefs they are in the name of the First Amendment about to protect, and perhaps the reinforcement and reaffirmation of the general norm, which rejects those beliefs and with which we are ourselves in accord. It is in both functions that the coercion occurs, for the judges' statements make clear that those who hold these views, and act on them in the ways that Collin's group was about to, are deserving of our reproach; the words of the judges themselves constitute a form of official censure and thus a kind of coercion and punishment, as well as the threat of other punishments.

That, however, is not necessarily to say the judges were wrong in saying what they did. It does point up the disjunction referred to at the beginning of the chapter, about the differential in our willingness to employ different *forms* of sanctions, primarily in the "legal" and the "nonlegal" categories, which is such a paradoxical and problematic element of our cultural attitudes about the appropriate responses to speech activity. It might well be asked, if we are so willing to apply nonlegal forms of coercion, then why not legal forms?

At least on the ultimate question whether legal coercion was constitutional, however, the courts were quite emphatic that it was not. They were less clear, on the other hand, on the score of whether this was a desirable result and of what precisely were the justifications for it. The opinions convey a strong sense of helplessness on the part of the judges. The dominant image suggested by the opinions is that of judges compelled to reach the results they did.

This sense of a predetermined result was created in several

ways. There was the usual invocation of "the First Amendment" itself, or "the United States Constitution," as if these words and texts constituted something firm and specific on the issue before the courts. The judges also seemed to intimate that they would arrive at a very different resolution if they were deciding the case on a clean slate, but that such a resolution was foreclosed by standing precedents of a higher court, the Supreme Court. Thus, the Illinois Supreme Court opened its opinion by declaring it was "bound by the pronouncements of the United States Supreme Court in its interpretation of the United States Constitution," pronouncements that "compel us to permit the demonstration as proposed, including the display of the swastika," and closed with the statement that the result had been reached "albeit reluctantly."[40] The federal court of appeals, after making its opening declaration of sympathy with the beliefs of the Skokie community, then changed ground rapidly and pronounced its own personal views as irrelevant to its assigned judicial task, which was described as applying the principles dictated by the Constitution: "As judges sworn to defend the Constitution, however, we cannot decide this or any case on that basis. Ideological tyranny, no matter how worthy its motivation, is forbidden as much to appointed judges as to elected legislators."[41] According to the court, its function was only "to decide whether the First Amendment protects the activity in which appellees wish to engage, not to render moral judgment on their views or tactics."[42] The form in which the court stated its conclusion, "[W]e find we are unable to deny that the activities in which the appellees wish to engage are within the protection of the First Amendment,"[43] at least demonstrates the utility of the negative form in our language as a means of conveying a certain reluctant state of mind and, additionally in this instance, the untouchable nature of the beliefs being protected.

This theme of predetermination was executed primarily through the methodology of analysis of precedent, a process involving the comparison of various details of this case with earlier decisions in order to determine, in a sense, what species was being handled. As is so often the situation, this turns out to be rather wooden process—at least in terms of its interest for fundamental theory—since once the characteristics of the relevant categories are treated as fixed, it then becomes only a matter of ascertaining how many of those characteristics are present in the case at hand. Whether the characteristics *ought* to be treated as relevant is a question usually left unaddressed. In the *Skokie* case, however, this uncritical process retained at least the interest of offering a virtually complete description of the First Amendment doctrinal architecture that has been built up during the last five decades.

The city structured its legal argument to locate this case within the established exceptions to the First Amendment rules, any one of which could independently have supported the city's position. It argued that the Nazi speech would constitute "fighting words,"which the Supreme Court had declared in *Chaplinsky v. New Hampshire*[44] to be unprotected, on the ground that certain types of speech lacked sufficient "social value" to justify protection. In *Chaplinsky* the state had convicted a member of the Jehovah's Witnesses sect for calling an arresting officer a "damned fascist," among other things. Such "fighting words," the Supreme Court held, were not within the doctrine of free speech, as was also true, it added, of libel, indecent language, and obscenity.[45] Naturally, in light of this, Skokie also argued that Collin's messages constituted "false statements of fact," which, like libel, were therefore unprotected. The city extended this line of argument by calling the speech "group libel," involving as it did the defamation of Jews, which in the 1952 case of *Beauharnais v. Illinois*[46] the Supreme Court had said could also be prohibited constitutionally. Indeed, Skokie had modeled its ordinance after the Illinois statute that *Beauharnais* had upheld.[47] Intervening decisions of the Court, however, cast considerable doubt on the continuing vitality of that case. Skokie further contended that what Collin and his group had to say was "obscene" and therefore, unprotected under established precedents dealing with pornography.[48].

Following on Holmes's test from *Schenck* and *Abrams*, the city claimed that there was a "clear and present danger" of a serious social harm likely to result from the march. The speech, it said, was especially offensive to the Jewish members of the community—in fact, the psychic equivalent of a physical assault.[49] Finally, the city argued that its regulation was not really a prohibition of speech at all, but only a regulation of its "time, place or manner," a category of regulation the Supreme Court had repeatedly held subject to a less stringent form of First Amendment review.[50]

As to all these claims, the courts found fatal flaws. The "fighting words" doctrine of *Chaplinsky* was said to apply only to certain personally insulting epithets spoken in a face-to-face encounter.[51] Here the speech, however obnoxious, was about "political ideas," and it could be avoided simply by not showing up at the village hall on the afternoon of the march. Similarly, as to the argument that this was libel, it was answered that these were not factual assertions, which could be gauged as true or untrue, but political ideas, as to which the Supreme Court had pronounced: "Under the First Amendment there is no such thing as a false idea."[52] The *Beauharnais* decision, upholding a group libel law that was not limited to factual falsehoods, was now of doubtful validity but, in any event, was distinguishable because it had involved a statute that had been interpreted to apply only to instances where there was a likelihood of violence; Skokie had by this time withdrawn its claim that the Skokie community was likely to respond violently to the Nazi march.[53] Nor could the Nazi speech be deemed "obscene," since it lacked the requisite quality of the erotic.[54]

The courts said further that this was not a case for the "clear and present danger" exception. According to more recent Supreme Court decisions, in particular the 1968 opinion of *Brandenburg v. Ohio*,[55] involving racist rhetoric at a Ku Klux Klan rally,[56] speech could be prohibited only on the basis of its dangerousness when it sought to incite others to serious unlawful behavior and persuasion was imminent. In Skokie it was, of course, implausible that anyone likely to listen to the Nazi speech would be immediately persuaded; the only violence that might arise would be from spectator hostility, which, again, the city had refrained from urging as a likely reality justifying prohibition and which, in any event, the Supreme Court had severely limited (perhaps even eliminated entirely) as a relevant consideration in free speech cases in various decisions denying "hecklers" any "veto" over unwanted speech.[57] Finally, as to the time, place, or manner claim, this went afoul of a now firm distinction between regulations that sought to limit speech because of the content of its messages and those that limited speech incidentally in the pursuit of other concerns; the exception applied only to the latter.[58] The Skokie regulation, which limited speech because of its harmful impact on, or its offensiveness to, others, was quite clearly directed at the content of the speech and therefore subject to the severe strictures of the First Amendment.[59]

And so it went with the courts' doctrinal interpretation. Obviously, such a method of treatment and analysis of the issues raised by the *Skokie* case did not provide any prescriptive justification for the result reached. It involved merely a description of the doors through which we must pass before we are clear of the citadel, and of the passwords needed to open those doors. It did not tell us why those doors are there in the first place—whether they have any positive function other than to prevent and channel passage. Wigmore's challenges, therefore, remain unmet.

In the *Skokie* opinions, however, one argument for protection, while also drawn from the precedents, yields an attempt at positive justification. Like the claims from precedent and higher authority, it is an argument that implicitly portrays the judges as being in a somewhat helpless position with respect to determining the outcome of the controversy. It was offered primarily as a response to the Skokie claim that prohibition of this speech should be permitted because of its deep offensiveness. But it might have been a response to any of the several reasons offered for suppression.

This justification was simply the inability to draw any line that would effectively exclude this kind of speech while not intruding on speech that everyone believed valuable and worthy of protection. Of all the arguments advanced in the *Skokie* case, that heard with the greatest frequency was this claim: to permit this speech to be restricted would jeopardize the entire structure of free speech rights that had been erected. According to the most commonly used illustration of this argument, to permit Skokie to ban this speech because of its offensiveness would mean that Southern whites could ban civil rights marches by blacks.

The principal case used to support this proposition was *Cohen v. California*,[60] in which the Court had said that California could not prohibit, on grounds of its offensiveness, a person from wearing in public a jacket inscribed across its back Fuck the Draft. To extend to the state the power to limit "indecent speech," said the Court, "would effectively empower a majority to silence dissidents simply as a matter of personal predilections."[61] No "readily ascertainable general principle exists" for drawing such lines, "[f]or, while the particular four-letter word being litigated here is perhaps more distasteful than others of its genre, it is nevertheless often true that one man's vulgarity is another's lyric."[62] Such statements as these became the bulwark of the decisions in *Skokie*: "The result we have reached," said the federal court of appeals, "is dictated by the fundamental proposition that if these civil rights are to remain vital for all, they must protect not only those society deems acceptable, but also those whose ideas it quite justifiably rejects and despises."[63]

According to this primary justification, protecting speech like that in *Skokie* is the necessary price of our own liberty. The abuse of freedom is inevitable and it must be tolerated along with the good in order to ensure that the good itself is tolerated. Collin and his collection were the flotsam of the First Amendment. We might regret it, but there was nothing we could safely do about it.

Yet, as confusing and multidimensional as *Skokie* was, a close reading of the opinions sometimes reveals intimations of some more positive social gain to be derived from the experience. At one point the seventh circuit said, somewhat elliptically, "It is, after all, in part the fact that our constitutional system protects minorities unpopular at a particular time or place from governmental harassment and intimidation, that distinguishes life in this country from life under the Third Reich."[64] Here, as in a few other passing comments, one can intuit some sense that the protection of this speech contains some deeper significance for us than that of incapacity to draw lines. There is the suggestion that we should be proud of this result, not just accepting of life's imperfections or even just glad that that acceptance protects us. But what might justify that feeling of pride, if indeed the feeling is appropriate at all, is never stated. Like just about everything in *Skokie*, so much is left unspoken that one is never sure about the meaning of what is said.

## III

We began this chapter by noting the extraordinary disjunction between our attitudes about the appropriate limits of tolerance for speech under the free speech doctrine and in our personal lives. It is the extremeness of the free speech principle, so highlighted when set beside our general behavior toward the speech acts of others, that seems so puzzling. Why should we pursue the idea of speech liberty to the degree we do? Why should the society so restrain itself that it must await the point where persuasion is about to bring consequences virtually all agree are to be avoided? Why should we force ourselves to ignore those who are ineffectual though equally intent on achieving the same consequences? Where else in life do we regard it as desirable to ignore that which is thought harmful until the final moments when the harm is about to occur? Solicitude for those bent on individual and social injury seems an odd virtue.

The challenges by Wigmore and by those who have walked in his path, are hardly addressed by the opinions in *Skokie*. One might have hoped for a full and rich appreciation of the gains to be had by protecting speech of that kind, a clear sense of the purpose or purposes behind the concept of free speech and of their meaning for the dispute, and a consciousness of the costs of pursuing those purposes at the expense of other possible aims. Instead, the reader was treated to a tight doctrinal analysis that avoided rather than confronted the social meaning of what was being done. When the judges peered out from behind the walls of doctrine, it was only to plead further helplessness through an argument about the inability to draw the distinctions required to decide the case the way they said they would personally have done.

Perhaps this is unfair. The line-drawing claim is a standard argument in law, going under such familiar appellations as "the slippery slope" and "the camel's nose in the tent." It rests on a number of premises, especially about the nature of language and human psychology. Words cannot always capture for us the sense we may have at the moment of some distinction we think it might be proper to draw. When we turn to open-ended, ambiguous words like *reasonable* or *dangerous*, we create the opportunity for distinctions to be drawn later that we did not originally intend. This may occur intentionally or inadvertently, but a wavering, blurred line can bring bad consequences even when applied by the well-intentioned. Differential application to similarly situated persons can be expected, leading to the sense of personal injustice and reduced respect for the legal system.

To recognize these considerations is only a beginning, for the fact is that in any legal system, and certainly in our own, these open-ended terms are inevitable. The choice we face is not between a legal system without the uncertainty of language and one with it. That is a false choice. Our present system of law has an abundance of the most abstract legal terms imaginable, and, what is of greatest significance here, that is true even in the area of free speech itself. Courts regularly decide who is a *public figure* in the libel field or what is *obscene* or what constitutes *commercial speech*—terms that do not seem substantially more constrained in meaning than do others that might be used, like *Nazism* or *genocide*.

The problem we face, then, is not only how much uncertainty a given legal rule will introduce into our law but also when we will choose to live with that uncertainty and when not. Sometimes our willingness will depend on the degree to which we think we have experience with the terminology, since that can give contour to otherwise meaningless jargon. But it would be nonsensical for us to avoid ever embarking on any new venture because of the absence of prior experience, though that seems to be the gist of the line-drawing argument in many cases. The important point is to see that the risks of uncertainty of lines is but one of many considerations in any given case, a consideration we sometimes give great weight to and sometimes not. We must therefore try to understand what lies behind this willingness to tolerate the risks of ambiguity in law, as well as to ascertain what exactly those risks are.

But I would go even further and suggest that the line-drawing claim is one of the most beguiling methods of obfuscation and diversion in legal argumentation, one that often serves as a convenient disguise for other purposes and motivations. In form and substance it is rather like what is known in law as a legal fiction, a statement that seems plausible in context but whose persuasive power is typically swollen by other, more compelling reasons that remain beneath the surface. The difference between the conventional legal fiction and the line-drawing claim is that the former usually involves simply calling something new by an old name, whereas the line-drawing argument usually has some actual relevance and can serve as an independent argument in its own right. One suspects that is why it has not received the same attention in the literature as legal fictions have.[65] But the critical point I wish to make is that the line-drawing claim is an appealing argument for any disputant, and especially for the free speech advocate, because of two primary characteristics: First, it shifts the burden of argument onto one's opponents; second, it seemingly reduces one's responsibility for the result being reached. There is much to recommend such a line of argument when one has only an intuitive sense that the result makes sense or when one has a justification clearly in mind but for other reasons is uncomfortable about admitting it openly.

This, one suspects, may well have been the case with *Skokie*. Given the premises or social reality offered in the *Skokie* opinions, it is difficult to believe that some workable rule could not have been arrived at for the speech at issue in that case. Speaking personally, I do not believe my own liberty of speech (or the speech I think it reasonable to value) would have been threatened by grafting such an exception onto the First Amendment, just as I do not now feel threatened by the constitutional dispensation for obscenity laws. Nor do I find it difficult to distinguish in my own mind between the type of "offense" caused by blacks marching in the South for their civil rights and that brought about by Nazis who would advocate the murder or enslavement of a segment of the community. (Perhaps one would conclude that the ideas advocated by the Nazis in Skokie were something different from this, and, if so, that would raise a separate question.)

Of course, such personal viewpoints ought not to be regarded as dispositive of the general issue; but they certainly are relevant and a worthwhile starting point, and by stating my own I hope to invite others to arrive at their own honest judgment. It seems a significant piece of corroborating evidence that virtually every other western democracy does draw such a distinction in their law; the United States stands virtually alone in the degree to which it has decided legally to tolerate racist rhetoric. This distinctive feature of American society in the world community is higlighted by the fact that the United States has yet to ratify either the Convention on the Prevention and Punishment of the

Crime of Genocide (which prohibits, among other things, the "direct and public incitement to commit genocide") or the International Convention on the Elimination of All Forms of Racial Discrimination (which prohibits, among other things, the "dissemination of ideas based on racial superiority or hatred"), in part because of concerns about potential conflicts between the conventions and the First Amendment.[66]

These are matters we must take up at length. Even if one accepts the line-drawing difficulties as relevant to the *Skokie* issue, other complementary reasons may better support the result, and it is obviously important to identify them if they exist. I have already suggested how the opinions intimate the existence of alternative rationales. It is a primary curiosity of the *Skokie* controversy that many free speech proponents saw (or at least proclaimed) the dispute as an "easy case," though there was also a good deal of despair in free speech quarters about the fickleness of people when it actually comes to implementing liberties and about the splitting up of the old alliance of liberal groups that had been forged during the 1960s.[67] Yet, at the same time the *Skokie* case is also noteworthy for the apparent inability of these proponents to translate their position into an articulable, coherent, and persuasive case.

## IV

To unravel these puzzles of the First Amendment will take some time and doing. But the puzzles we face go beyond the one I have focused on in the discussion thus far. I began by raising the problem of the disjunction between our attitudes about the limits of nonlegal coercion toward speech behavior and our attitudes about the limits of legal coercion toward the same behavior. In the process of focusing attention in that way on the contemporary application of the free speech concept, we have been led to ask whether it makes any sense for the society to commit itself to what can only reasonably be regarded as an extreme extension of the free speech idea. This issue of the proper limits of free speech really only touches one dimension of a more general problem of developing a basic free speech theory.

At least two other major questions must be addressed if we are to contribute to the development of such a theory. First, does it makes sense to treat this one area of behavior—that is, speech behavior—differently from other areas of behavior—that is, nonspeech (or nonverbal) behavior? Not only does there appear to be a disjunction between our nonlegal and legal responses to speech activity but also between our legal responses to speech and nonspeech activity within the society. As with umpires in the game of baseball, we are expected to tolerate an enormous variety of verbal abuse from incensed managers, yet not the slightest physical contact. The question to be considered, then, is whether this second area of apparent incongruity in the web of social rules can be justified.

A last major theoretical issue is whether it is sensible to place the authority over the interpretation and enforcement of a free speech principle in the hands of the judiciary. One might very well conclude that a society like ours ought to grant special protective status to speech activity and yet decide that the implementation of that principle should be through the usual processes of democratic decision making—which is, of course, exactly the approach followed in Great Britain. Yet, in our society that choice has not been made; instead, an unelected body is vested with the power to interpret and apply the free speech principle and to override legislative actions that abridge it.

Branching out from these fundamental theoretical issues are, of course, a host of others, especially labyrinthine doctrinal questions about application of the free speech principle to particular factual contexts. These matters cannot possibly be comprehensively examined in a book of this kind, but the basic vision of the First Amendment—the central aims and goals we seek to achieve through its operation—is absolutely critical to any analysis of those particular problems, and helping to define that vision *is* the purpose of this book. Still, a central premise of the inquiry undertaken here is that a special focus on the issue of the extremes to which the First Amendment has been pursued in this century will help us grasp a general vision of the amendment. From that vantage point we should be in a good position to gain insight into the other major fundamental issues of free speech—the questions about the differential treatment of speech and nonspeech behavior and about the legitimacy or desirability of judicial review in the free speech context.

How shall we proceed with the inquiry? In the next two chapters I shall try to isolate the major obstacles encountered in any attempt to build a free speech theory out of the conventional building blocks of contemporary thinking about free speech. I divide that thinking into two primary categories and reserve one chapter for each. The first approach to free speech, which I refer to as the classical model, is a brief summary of what people tend to say and think generally in public discourse in defense of the free speech principle. The second approach, which I refer to as the fortress model (chapter 3), tends for reasons that will emerge later to play a more submerged role in our public discussion; while more difficult to identify as a fundamental way of thinking about free speech questions, it is nonetheless of at least equal importance to that associated with the classical model.

For several reasons I begin with this discussion of the existing, or present, state of thinking about free speech rather than plunge immediately into the perspective I wish to offer (which is saved until chapter 4). Simply as a matter of organization, it seems sensible to consider first what the existing approaches have to offer on the question we are addressing. I have criticized the opinions in *Skokie* for their vagueness about the reasons for the results reached in that case. Perhaps, however, this is only an idiosyncratic and not a general failing; perhaps the theoretical structure of the First Amendment, however amorphous and fluid it may be, provides convincing reasons for the results, but the judges there just somehow failed to articulate them. This would not be quite so damning a criticism of those judges as it might at first appear. Judges, after all, are generalists in function, as much by design as by necessity, and very often the considered thought of those who specialize in the field fails for years, even decades, to make its way into the corridors of practical decision making. On the other hand, it is also entirely possible that the theoretical foundation of free speech has failed the judges, has broken down at the point where it was needed in the field, causing the confusion exemplified by a case like *Skokie*. Which of these is true is the issue we now pursue.

There are other reasons behind the organizational scheme I have chosen. We need, as we so often do, to trace the general outlines of our existing thought patterns to be in a better position to appreciate the significance of a different perspective. This is especially true when, as here, the pieces of the new perspective already exist in the crevices of our present methods of thinking and, especially, in the often overlooked criticisms to those methods. Ultimately, I believe, it is only by feeling the nature, and indeed the justice, of the desire to punish speech acts that we will ever be able to arrive at a true understanding of the social benefits potentially to be derived from tolerating that behavior.

# 2
# The Classical Model and Its Limits

When we are deciding whether some rule should govern a dispute, especially if the dispute is thought to be a borderline case, a familiar and sound injunction is that before venturing an answer, we should first know what the basic purposes are behind the rule. Despite both its familiarity and its soundness, however, the injunction is frequently ignored. But our neglect makes it no less valid. Without knowing what we are trying to achieve by a particular rule, we cannot sensibly proceed to offer an intelligent opinion about which facts from the mass of worldly detail are important and which irrelevant, about how much weight ought to be given to the facts that are deemed relevant and about how the various elements of the problem ought finally to be resolved in a judgment—however inscrutable the process of rendering that "judgment" might seem to, or actually, be.

So it is with free speech. A case like *Skokie* requires us to refer back to fundamentals. If, as it was frequently argued and tacitly assumed to be, *Skokie* was a case at the margins of the First Amendment, then we need to know something about the "core" of the First Amendment, about its underlying suppositions and premises, about what we seek to accomplish by the amendment, before we can decide what to do at any point along the periphery. Under our inherited ways of thinking about free speech—what in this chapter I refer to as the classical model—that core is conceived to be the speech we truly value.

That is to say, the logic of the classical vision of the First Amendment takes the following form: Since it is "speech" we are protecting under the free speech principle, it must be because some speech is of great importance to us. It follows that any "theory" of the First Amendment must proceed from an inquiry into why having the liberty to speak is valuable in the first place. Contemporary analysis of free speech follows this theoretical path, but theorists disagree about which identifiable "values" ought to be given precedence over others.

In this chapter we begin by examining what over time people have identified as the principal values of speech. Then we will be in a good position to say, first, how much value from a free speech perspective there is in a case like *Skokie* and, second, whether that positive value is sufficient to outweigh in our minds the individual and social harm caused by that speech.

We consider here the essential, elemental characteristics of the traditional, libertarian vision of freedom of speech. We seek an overview of the structure of thought that people bring to controversies over the regulation of speech, and in particular that they bring to cases like *Skokie*. We deal with the first line of defense of free speech—with the response you are likely to get if you asked the question Why do we have a principle of free speech in this society?

The classical vision of free speech has antecedents stretching far back in time. The primary connection is with the period of the Enlightenment, in the eighteenth century, when the interest and faith in man's powers of reason flourished and when there occurred that enormously important revolution in the way people conceived of the relationship between the state and the individual members of the society.[1] Two cardinal premises about social organizations arose from this transformation in thought: first, that the government is possessed of only limited political powers, which it derives from the citizenry; second, that the people themselves, as the ultimate sovereign, are competent to determine their own destinies. Out of these premises there arose that momentous political experiment, the American Revolution, which scholars like Bernard Bailyn describe as having "reconceiv[ed] the fundamentals of government and of society's relation to government."[2]

How do these reconceived notions match our contemporary realities? How well does the rhetoric of the preceding two centuries describe what is occurring today in the realm of freedom of speech?

## I

In today's discourse about free speech, the dominant value associated with speech is its role in getting at the truth, or the advancement of knowledge.[3] Speech is the means by which people convey information and ideas, by which they communicate viewpoints and propositions and hypotheses, which can then be tested against the speech of others. Through the process of open discussion we find out what we ourselves think and are then able to compare that with what others think on the same issues. The end result of this process, we hope, is that we will arrive at as close an approximation of the truth as we can.

This idea is so familiar to us that we may overlook its significance. Other benefits of speech can be, and sometimes are, recommended. Speech, it is sometimes said in contravention to the truth-seeking conception of free speech, should be specially protected because it plays a critical role for each individual in achieving self-fulfillment through the act of self-expression, and not because of the practical benefits it yields in the way of uncovering truths about the world; or because our basic social morality dictates that every member of society be treated as an equal, with dignity and respect.[4] Still, the practical benefits of gathering information and ideas for the truth-building process continue to be the major attraction claimed for speech and hence the basis of our contemporary theory for the free speech principle.

In recent decades, however, the value of speech for seeking truth has been thought about in a narrower context—that is, in the political context of democratic self-government. In a democracy, it is commonly said, speech performs an essential role in assisting the citizens in their collective search for answers to the issues they face. Professor Zechariah Chafee, the first major American scholar of the First Amendment, wrote in his book *Free Speech in the United States*[5] that the First Amendment protects "two kinds of interests"; one is "a social interest in the attainment of truth, so that the country may not only adopt the wisest course of action but carry it out in the wisest way," and the other is an "individual interest, the need of many men to express their opinions on matters vital to them if life is to be worth living."[6] The individual interest Chafee identified has played a secondary role to the social interest he mentioned, and the social interest has come to mean primarily a political interest.

One of the seminal works on the value of freedom of speech for a democratic society is an essay written in 1948 by the philosopher and educator Alexander Meiklejohn. Entitled simply "Free Speech and Its Relation to Self-Government,"[7] Meiklejohn's essay developed what to many has seemed a self-evident but nonetheless highly significant thesis: The principle of free speech plays a practical role for a self-governing society, protecting discussion among the citizens so that they can best decide what to do about the issues brought before them for decision. If one has chosen to live in a self-governing society, Meiklejohn argued, it follows as a matter of course that the government is, and ought to be, forbidden from interfering with any speech by the citizens that is concerned with the performance of their sovereign functions. A *Robert's Rules of Order* may be necessary and appropriate, but beyond that the citizens can brook no further interference into their political dialogue.

As his paradigm for thinking about the functions of free speech, Meiklejohn (like so many others) took the traditional New England town meeting, at which the citizens regularly met "as political equals." Order was imposed by the moderator, but no idea as such could be suppressed. This was because the citizens, as ultimate sovereigns, needed to make their own decisions and the best decisions they could: "Now in that method of political self-government," Meiklejohn wrote, "the point of ultimate interest is not the words of the speakers, but the minds of the hearers. The final aim of the meeting is the voting of wise decisions. The welfare of the community requires that those who decide issues shall understand them. They must know what they are voting about. And this, in turn, requires that so far as time allows, all facts and interests relevant to the problem shall be fully and fairly presented to the meeting." In this setting, all viewpoints must be given an opportunity to be heard: "Just so far as, at any point, the citizens who are to decide an issue are denied acquaintance with information or opinion or doubt or disbelief or criticism which is relevant to that issue, just so far the result must be ill-considered, ill-balanced planning for the general good. *It is that mutilation of the thinking process of the community against which the First Amendment to the Constitution is directed.*"[8]

For Meiklejohn, then, as for Chafee, it appeared to be the relationship of speech to the results, to the outcomes of the deliberations, that made speech activity worthy of special protections. Because all citizens in the community share an interest in the results, the protection of speech serves a "collective" interest and not, as Chafee had added, any single individual interest in self-expression. In the final analysis, "[w]hat is essential is not that everyone shall speak, but that everything worth saying shall be said."[9]

Under his theory Meiklejohn limited the scope of protection afforded by the First Amendment to "political" (or, as he referred to it, "public") expression.[10] This limitation was not quite so severe as it at first seemed, for when many objected that much valuable speech (like artistic expression) would go unprotected under his theory, Meiklejohn acknowledged the importance of such speech and denied that his theory precluded protection of it. Since it was "relevant" to political decision making, it could therefore be included in the category of "political" speech. Knowledge of Shakespeare is as essential to wise political decisions as any explicitly political expression, Meiklejohn congenially rejoined.[11]

Meiklejohn's ideas about how the political functions of speech justify and define the concept of freedom of speech have become a major part of the modern idiom of the First Amendment. We now hear regularly that speech is an essential tool to a self-governing society. A commitment to democracy, it is said, entails within it an agreement that governmental power to deprive citizens of the opportunity to discuss openly all public issues must be sharply restricted. Open discussion is touted for its role in improving the quality of decisions. Practical, concrete benefits are said to flow to the political community from this process, and it is these collective social benefits that we protect through the First Amendment. A recognition of the primacy of political dialogue has emerged in the thinking and judicial decisions regarding the free speech principle.[12]

The principal Supreme Court decision that most embodies these features of the modern First Amendment is *New York Times Co. v. Sullivan.*[13] At issue in *Sullivan* was the constitutionality of state common law rules relating to defamation—at least as applied to suits brought by public officials who seek damages against the press for false statements made about them. In deciding whether to apply constitutional strictures to these common law actions, the Court worked from the basic premise that in a democracy the "people, not the government, possess absolute sovereignty," in the words of James Madison.[14] From this premise the Court reasoned that the government had no business interfering with, or seeking to limit, public criticism by citizens of their elected representatives. The "central meaning of the First Amendment," the Court thus pronounced, must be that a regime of seditious libel, under which citizens could be punished for exercising their sovereign right to criticize the government, was unconstitutional.[15] Instead of trusting the citizens to exercise their political prerogatives, in such a regime the state had usurped these functions for itself.

From this notion of the core meaning of free speech, the Court was able to approach the issue of common law defamation rules by asking whether they were sufficiently analogous to the seditious libel regime to warrant constitutional trimming. To this question the Court answered with a qualified yes, concluding that defamatory statements of fact must be protected to some extent (absent knowledge of falsity or reckless disregard of the truth) in order to shelter the legitimate interests of citizens in exercising their sovereign governmental powers. The Court saw virtue in creating an environment in which speech was "uninhibited, robust, and wide-open."[16]

*Sullivan* was a landmark decision for the First Amendment. In a famous article on the *Sullivan* opinion that appeared the same year, Professor Harry Kalven pointed out how the Court had for the first time given the First Amendment a theoretical starting point in the finding that the "central meaning of the First Amendment" was the disallowance of any regime of seditious libel.[17] The "clear and present danger" test—whatever one might think of it as a test—had been theoretically empty.[18] What *Sullivan*, did, therefore, Kalven pointed out, was to tie the meaning and function of the free speech principle to the structure of the political system, identifying its vital role in a democratic system of government. In doing this, of course, the Court had essentially adopted the Meiklejohn approach, Kalven pointed out.[19] Ironically, the *Sullivan* opinion nowhere mentioned Meiklejohn, but Kalven's article confirmed the association and it thereafter became axiomatic.[20]

This Meiklejohn-*Sullivan* alliance, with its heavy emphasis on the practical importance of freedom of speech for a democracy, can be traced through many subsequent First Amendment cases.[21] It provides the core structure around which much First Amendment discourse and many opinions are built. Because it has achieved a kind of ascendancy as a free speech theory, we must give it our close attention. But once we do, we quickly discover that the relationship between a free speech principle and a self-governing society is more ambiguous then we commonly suppose or are told. Once we have sorted out that ambiguity, we can then begin to evaluate to what extent the identified virtues of speech justify a free speech principle as we have come to know it.

## II

The Meiklejohn-*Sullivan* perspective on the central meaning of the first amendment harbors at its roots a profound confusion. The difficulty rests in its assumption that a commitment to the democratic system of government entails a comparable unqualified commitment to a free speech principle as we know it. The error consists in taking a point that is valid in one context into another context where it may well have no, or at best only limited, explanatory power.

It is quite correct to argue, as Meiklejohn and *Sullivan* did, that a self-governing society cannot tolerate a government that seeks to usurp the sovereign decision making authority of the citizens by denying them the opportunity to hear ideas and viewpoints. But that is not the same thing as saying that we, the citizens, must in a self-governing society agree, without more, that *we* cannot choose to limit and regulate speech activity within the society and remain a self-governing community. A regime of seditious libel imposed by the government, acting independently of the democratic procedures, as Kalven noted, "strikes at the very heart of [a] democracy." "Political freedom ends when government can use its powers and its courts to silence its critics."[22] But if the people themselves, acting after full and open discussion, decide *in accordance with democratic procedures* that some speech will no longer be tolerated, then it is not "the government" that is depriving "us," the citizens, of our freedom to choose but *we* as citizens deciding what the rules of conduct within the community will be. Then the "democracy" has functioned, and it may be asked whether it does not "[strike] at the very heart of [a] democracy" to say that the citizens cannot choose to make that decision.

It may well be true that there is a point in the regulation of speech activity at which "we," as self-governing citizens, may strip ourselves (or a part of the community) of the capacity to exercise self-government in any realistic sense; then it would be correct to say to us that our commitment to self-government entails a limit on our freedom to engage in such an act of political self-dispossession. The point now has coherence. Just as we might properly claim to be defending liberty if we stop a man from exercising his liberty to sell himself into slavery, so might we properly claim to be defending democracy if we stop a majority from voting to convey upon the government the power to punish all public criticism of the state.[23]

This caveat does not take one very far, or at least not far enough to account for most of what has been happening under the banner of free speech in this century. Even if one concedes the essential validity of the argument just presented (as being either coherent or persuasive), it will justify intervention into the democratic decision making relating to speech activity only in order to preserve the minimally essential conditions of a "democratic" society. That, of course, is itself a highly ambiguous standard, but it would seem reasonably clear that few if any of the restrictions on speech we have encountered over the last sixty years, and the rejection of which now form the basis of our First Amendment jurisprudence, could be fairly described as jeopardizing the elemental structure of a democracy—or, stated another way, that the absence of these regulations was the sine qua non of a democratic political system.

If, then, we ask what "we" as citizens in a political community committed to a democratic process would choose to take as the limits on "our" power to regulate speech activity within the society, we might have to accept the minimal limit we have just noted and we might choose to accept more. But our acceptance of that greater limit would certainly not automatically be compelled by the fact that we had already committed ourselves to resolve issues between us according to a democratic norm of majority vote.

Here, however, a point must be emphasized. Thus far we have concluded that to commit ourselves to democracy, without more, is at most to commit ourselves to a free speech principle that is strong enough to curtail governmental interference with speech activity when that interference is designed to usurp the sovereign prerogatives of the citizenry, and to refrain from regulating speech ourselves to such an extent that self-government is no longer reasonably feasible. But this does not mean that we may not appropriately *choose* to structure our political and social system in such a way as to put into effect a more vigorous and thoroughgoing free speech principle. We may decide, as indeed most of us would probably be eager to do, that our "democracy" will mean something more than a raw principle that majority rule on all matters shall prevail. That choice, however, must be explained by more than a reference to our commitment to "self-government" and to examples where the state is reclaiming the autocratic political authority of the pre-Revolution English kings. It seriously overloads the term *democracy* with unstated and unclear meanings to give it greater content, as happens in opinions like *Sullivan*.

We can agree, then, that speech, even highly extremist speech, ought to be "protected" under a free speech principle from restriction by a government that is seeking to deprive the political community of the chance to decide itself what to do about the speech. This agreement derives from our commitment to a democratic political system. We have an interest in preserving the *opportunity* to exercise self-governance, but this is not the limit of the free speech principle we have developed or are now attempting to explain. The First Amendment has not been confined to imposing limits on errant, undemocratic, official efforts to control speech but on democratically sponsored efforts as well.

It is *that* limit that must be justified but cannot by simple reference to a preexisting choice for democracy.

Courts and writers too often end their analysis at the point where they have proclaimed the function of free speech to be to stop "the government" from interfering with the sovereign authority of "the people," and the reader is left with the image of unrestrained governmental power being checked by the courts. Such a portrait usually falsifies the reality, masking a deeper conflict within the political community itself, out of which the regulation at issue emerged with the support of a majority of the citizens. Yet, courts have not considered it relevant in free speech cases to inquire into the democratic pedigree of the regulations at issue, and then extend their approval to those shown to have the requisite papers—as both *Sullivan* and *Skokie* demonstrate. We shall later consider why courts tend to portray the world as other than it usually is in this regard (or, at least, stop considerably short of describing it accurately), including their role in it. For now, however, we bypass that question and continue considering whether there is any good reason to be found in the classicial model why we should accept the limitations imposed on our own regulatory powers by the free speech principle.

Let us focus our attention on the principal free speech "value" already identified: our practical interest in reaching the truth or wise decisions. (In the end, the same questions can, and will, be raised about the explanatory power of the self-fulfillment rationale.) Whether one defines the truth-seeking enterprise broadly or narrowly (as limited to the political arena), the advantages of free speech to that enterprise are said to justify extending special protection to speech activity. How and when are these benefits likely to be realized?

There is obvious validity to the general idea. One need not go through the elaborate trial of argument, from Socrates to

John Stuart Mill to Holmes and on to the present, to concede that few if any of us believe we are able single-handedly to arrive at the truth. From personal experience, all of us can draw on numerous instances where the process of open discussion advanced our understanding and led us to abandon falsehoods we once held as firm truths. From those experiences alone we should be, and are, disposed to recognize that toleration of at least some speech is essential to improve our chances of reaching the best possible judgments on the issues we are called upon to decide.

This claim must not be pushed too far. It is possible, of course, to argue, as Mill did, that since we can never know for sure that what we believe is true, we ought to be prepared to listen to everything, however convinced we are of its error.[24] But this is unacceptable. In a case such as *Skokie*, the chance that the Nazi messages may turn out to be "true" is hardly a persuasive basis on which to defend such speech, and few if any free speech advocates turned to this kind of argument in that context. The more we believe in the immorality or error of the ideas being expressed through the speech, the more attenuated is the truth-seeking advantage claimed as the justification for the free speech principle. The "value" to us in these terms ranges from remote to none. Just as in libel area the Court has sometimes recognized no "value" in defamatory false statements of *fact*, so we might appropriately extend that to at least some portion of the realm of opinion as well.

Suppose in deciding on the merits of toleration we begin with the assumption not that what is being said may be true but rather that it is false. Are there then positive benefits from a truth-seeking perspective that support protection? Here we encounter two primary arguments for speech protection.

First is Mill's major alternative argument for toleration. Through confrontation with falsehood, Mill argued, people retain a "livelier" sense of the truths they themselves already hold but which may have become stagnant: Through censorship, Mill claimed, we "lose, what is almost as great a benefit [as truth], the clearer perception and livelier impression of truth produced by its collision with error."[25] Truth requires regular exercise, as it were, and without it it atrophies into dogma.

A second claim for protection of false or bad speech focuses not on the benefits to us from interaction with it but rather on an important informational gain we may acquire simply by hearing the falsehood expressed. Meiklejohn's essay can be read to make this kind of argument, though it is by no means explicit.[26] It is more readily found in Chafee's *Free Speech in the United States.*[27] In any event, the basic idea is that by listening to extremist speech, we benefit simply by becoming aware of the presence within the society of this kind of dissatisfaction and dissent. To have such knowledge can be important to us in our policymaking role because we may want to take action—by improving educational and employment opportunities, housing conditions, and so on—to allay the underlying grievances we think may have given rise to the dissent. It is also usually preferable to have these potentially disruptive individuals and groups operating in public rather than in private, where the unwanted weeds of frustration and revolt may grow more rapidly from inattention and where the falsehoods being propagated may less easily be exposed for their error. A policy of near complete openness to speech, in this sense, provides us with a social thermometer for registering the presence of disease within the body politic and the best opportunity of administering a speedy cure.

As before, we must acknowledge the validity of these arguments as general propositions. While they are generally valid, whether or not any of these benefits will accrue sufficiently in a particular case, or even a group of cases, is at the very least an open question. One may seriously doubt, for example, whether it is always true that open confrontation with falsehood yields, or is necessary to, a richer belief in truth. Are the uninhibited activities of groups like the Nazis really that important to maintaining a vigorous belief that what they have to say is immoral and wrong? It seems an equally plausible theory as to some ideas, at least, that to regard them as unspeakable is the best method of rejection. Like all human activities, dialogue is not invariably useful under all conditions.

To emphasize this and other points, we might look at our behavior in another area of social life where speech activities pose difficulties in a context in which truth seeking is also a primary objective. Take the system of criminal justice and the regulations that govern jury trials. Here, where ascertaining truth is a basic aim of the process, our rules recognize both how speech is important to that primary goal *and* how speech has the potential to thwart that goal. We secure the right of the parties to present their respective arguments on the issues to be decided. But they may not speak in any way they wish. We take people as they are, and the system accordingly assumes that the process of rationality we prize so heavily may be undermined by speech itself, in particular by speech that is "inflammatory" or that may produce an improper "emotional" response.[28] Some evidence will be excluded because of its potential for prejudicing the jurors' judgment, because they will be inclined to give it more weight than it deserves. Furthermore, we impose various limitations on the manner of presenting arguments in the court, because respect for the process itself, or the individuals involved in that process, is also regarded as a prime value.

Such rules for governing speech in the forum of the criminal jury trial are suggestive, if not indicative, of the range of considerations that must be taken into account in deciding what limits there ought to be on speech activity. This is not to suggest that the particular resolution of the problem of limits that we observe there ought to be repeated in the general political forum, with which the First Amendment is primarily concerned. But the comparison certainly forces on our attention the need for a more extensive explanation of speech protection than the simple claim for truth seeking offers. In fact, just as with the argument from self-government, it may be argued with considerable force that an interest in seeking truth cuts against rather than in favor of speech protection in cases like *Skokie* and other extremist cases where, for example, the speech seeks to subvert the truth-seeking process itself. At the extremes, the values associated with the "core" speech become proportionately attenuated. Yes, we agree (as we imagine in our minds an Isaac Newton, a Thomas Paine, a Henry Moore), speech and discussion are vital to the search for truth, to the operation of democracy, and to individual self-expression, and we do value those ends greatly. But as the nature of the speech we consider moves progressively away from these most admirable kinds of expression, these great values associated with speech lose much of their force and may even be threatened by the speech itself, so that, now, to subscribe to those values seems to compel prohibition intead of protection.

Thus, we sec that at the same time there is an attenuation in the benefits associated with speech in the extremist speech case, there can simultaneously occur an escalation in the potential harm from the speech itself. The invocation of the generally accepted values of rationality and self-government and self-expression no longer work so smoothly; in fact, their invocation cuts in a sharply divergent direction on the issue of tolerance or intolerance of the speech activity. These thoughts are variations of the classic paradox for free speech (noted in chapter 1): a commitment to a principle of free speech can lead to protection of those who would advocate the abolition of free speech itself. And these thoughts were probably behind Wigmore's incredulity at the idea that in a democracy we should permit activities designed and intended to interfere with and upset the legitimately reached decisions of that system, by means other than through democratic procedures. As we have now seen, the same paradox can be found with respect to arguments for free speech based on its importance to individual autonomy and truth seeking: should a commitment to either purpose result in the protection of those who seek to destroy those values?

We need not rely on paradox alone, or deny altogether the potential for any of the benefits normally associated with speech activity, in order to feel that a persuasive case has yet to be made for the protection of such speech. The full question is not simply whether some benefits may be realized through toleration but whether we think those benefits are sufficiently great to outweigh the competing harms we will suffer if we choose that course. It is possible to imagine several advantages (even of the kind generally associated with exposure to speech activity) from being subjected to robbery, but their acquisition would hardly induce us to take up a banner for the right to rob. We must, therefore, now turn to the problem of assessing the harm from speech, especially speech of the kind in the *Skokie* case.

## III

The whole subject of the harm speech can cause is curiously dealt with under the classical free speech model. After a close and disinterested study of it, it is hard not to find truth in Wigmore's general accusation that free speech proponents frequently seriously underestimate the costs of tolerance of certain speech, especially extremist speech.

At the outset of this chapter, it was noted that the historical antecedents of contemporary libertarian theory regarding freedom of speech are primarily to be found in the Enlightenment period of the eighteenth century, when belief in the power of rationality, especially of the "common man," was widely proclaimed. Out of this development of the assumption of universal, inherent rationality the movement for democracy arose. It justified granting citizens the power of sovereignty. Additionally, it shaped how libertarians conceived of the harmfulness of speech.

When Milton argued for repeal of the British licensing system in 1644, his appeal was to a trust in the power of reason. He denied that falsehoods would ever prevail over truths, charging that to believe otherwise was an insult to Truth: "And though all the winds of doctrine were let loose to play upon the earth, so Truth be in the field, we do injuriously by licensing and prohibiting to misdoubt her strength. Let her and Falsehood grapple; who ever knew Truth put to the worse, in a free and open encounter?"[29]

Milton's statement, taken from another era and out of its context, may sound naive and unduly optimistic, but it reflects a widespread attitude reflected throughout Western liberal thought and encountered in contemporary discussions about free speech issues. Speech, especially speech that advocates an idea, is only a prelude to action and really causes no harm. Only "acts" bring harm, not "words"—as the sticks-and-stones refrain of our childhood proclaims. Speech is either entirely innocuous or will be stopped naturally from causing harm when confronted by the truth. If words near action, which, of course, may produce injury, we may then properly intervene to protect ourselves—as the "clear and present danger" test allows. Holmes's opinion in *Abrams* can be read to offer a repetition of Milton's optimistic forecast:

> But when men have realized that time has upset many fighting faiths, they may come to believe even more than they believe the very foundations of their own conduct that the ultimate good desired is better reached by free trade in ideas—that the best test of truth is the power of the thought to get itself accepted in the competition of the market, and that truth is the only ground upon which their wishes safely can be carried out.[30]

If we may borrow a metaphor from the Enlightenment figure Adam Smith, Holmes may be read as suggesting the presence of an unseen hand that guides truth to victory over the challenge of falsity.

To the extent that contemporary general thinking considers speech capable of producing harm, it identifies two potential risks: Speech may persuade people to do things that are socially undesirable and it may offend people who find the ideas expressed objectionable. As for the risk of persuasion, Milton's and Holmes's proposal that truth will naturally emerge victorious, if true, means there is little reason to be concerned. Often in free speech cases it will also be pointed out how unlikely it was that the speaker would persuade anyone in the immediate audience. Recall how Holmes referred to the *Abrams* defendants as mere "puny anonymities,"[31] suggesting that any concern over the dangerousness of these individuals was grossly exaggerated, even preposterous. "Nobody can suppose," Holmes thought it relevant to argue, "that the surreptitious publishing of a silly leaflet by an unknown man, without more, would present any immediate danger that its opinions would hinder the success of the government arms or have any appreciable tendency to do so."[32] The social consequences of the defendant's act were reduced to its potential to shape the behavior of others in the immediate context of the speech act. "[W]hatever may be thought of the redundant discourse before us it had no chance of starting a present conflagration,"[33] Holmes concluded in a later case.

Even when persuasion is a real possibility, it is commonly said, still other means are at our disposal short of outright prohibition and punishment for counteracting the risk: Answer with your own arguments, it is said. We can trust our citizens to distinguish right from wrong, truth from error. To permit prohibition of this speech is to underestimate their skills, as well as to deny them the opportunity to be educated through the use of those skills. Faith in the average citizen, or in the collective body of them, is a catechismal answer of the classical libertarian vision.

As for the offense suffered by listeners, another line of argument is advanced that reduces the asserted harm to inconsequential proportions. "Don't listen" or "Avert your eyes" is generally recommended.[34] In *Skokie*, for example, free speech proponents regularly pointed out that those individuals who were likely to be offended by the Nazi message could easily avoid becoming upset by taking advantage of readily available means of self-help—they could stay at home or enjoy a picnic in the countryside or simply do what they would have done anyway, which in all probability would have kept them out of earshot and eyesight of the events at the village hall. Since one need not suffer, the federal court of appeals appeared to reason, it seemed unnecessary to take into account the suffering it was claimed the speech would cause.[35]

Whether explicitly or implicitly, the classical defense of free speech views actions, but not words, as capable of inflicting cognizable injury.[36] When injurious conduct occurs, the society, of course, may properly intervene and take appropriate measures to protect itself. But until that point is reached, nothing really has "happened" and so the society possesses no legitimate basis for intervention. It is also, it may be noted here, this perceived differential in the harm-producing capacity of conduct as opposed to speech that is commonly seized upon as a justification for the differential treatment accorded the two areas of behavior.[37]

If we are to understand what harm speech can cause (or, to put it more accurately, what harm is potentially caused by prohibiting intolerance of speech activity), we really must know something about what drives people to want to suppress speech in the first place. Even if we assume, as the classical perspective does, that speech involves only a "risk" of something else that is truly harmful, we ought to know something about what people actually perceive to be the risk and what kind of effects there are on people who are exposed to that risk. In short, we should try to understand why people want, or need, to suppress speech acts through the operation of law.

Investigation of the modern free speech literature for thoughts about the origins of the need to suppress speech yields surprisingly little. There are frequent references to the emotion of fear, when the discussion turns to describing the motivations of those who seek to punish speech. "Those who won our independence by revolution were not cowards,"[38] Brandeis wrote in the *Whitney* case, saying by implication that those who seek to suppress speech are. "To be afraid of ideas, any idea, is to be unfit for self-government,"[39] Meiklejohn asserted. If fear motivates the desire for intolerance, then fear of what? The question is seldom answered.

We do, however, have one well-known statement on the need for legal suppression, again from Holmes, which he offered in his *Abrams* dissent. Just before his famous marketplace-of-ideas theory of free speech, Holmes spoke of how "[p]ersecution for the expression of opinions [is] perfectly logical." In his remarkable and pithy style, Holmes explained the logic: "if you have no doubt of your premises or your power and want a certain result with all your heart, you naturally express your wishes in law and sweep away all opposition. To allow opposition by speech seems to indicate that you think the speech impotent, as when a man says that he has squared the circle, or that you do not care wholeheartedly for the result, or that you doubt either your power or your premises."[40] Such a view, Holmes added, is shortsighted, at which point he made his celebrated recommendation that we place our faith in the "competition of the market."

Holmes's remarks on the need to restrict speech offer a useful point of departure for any analysis of the nature of censorship. He confirms, though by the rather odd use of the term *logical* to describe the desire for "persecution," the naturalness of the impulse to prohibit certain speech. And he, like others before him, notably Mill, centers that impulse on the nature of human "beliefs."[41] The impulse flows from the desire to protect those beliefs against contrary ones.

Although Holmes's remarks are readily seen as interesting and revealing, their real significance may not at first be apparent. Of special importance is his characterization of censorship as fundamentally a *form of expression*—actually, *speech*—that counteracts the messages one would otherwise naturally communicate through toleration or passivity; for "to allow opposition by speech seems to indicate" the series of messages he describes.

The act of speech, therefore, puts those who know of it, and who believe differently, in a serious dilemma. For them it is not simply a matter of choosing not to be offended—perhaps by "averting the eyes"—or a matter of standing like some outsider who is observing a process of potential persuasion at work. The dilemma they face is significantly more complex, for by doing nothing—by being tolerant—they may be contributing to the success of the beliefs they dislike. They are now, like it or not, part of a dynamic process, from which they can withdraw only at the risk of furthering that which they oppose. And that is why, as Holmes tells us, people in this situation are "naturally" inclined to "express [their] wishes in law and sweep away all opposition."

Both tolerance and intolerance, therefore, are *communicative acts*—speech in the simplest and purest sense—that spring from the need to make one's position in the world clear. The point, however, is rarely appreciated. Yet, to see this dilemma that arises from the communicative significance of each posture helps us to better understand the strength of the need for prohibiting certain speech activity. Given the fact that free speech theory itself urges us to recognize the tremendous need of every individual to express himself or herself, we can hardly ignore or minimize the power of that need just because it arises in the form of restraints on the speech of others.

Even this rich insight into the origins of the need to restrain speech activity, however, raises some puzzling problems. If we see intolerance as a form of expression intended to avoid creating the wrong impression—either that we don't really believe what we claim to believe or that we don't have the courage of our convictions or the power to defend them—we might well wonder why the desire should be so strong, if, as Holmes says, "you have no doubt of your premises or your power." Holmes's vision of intolerance postulates a relatively fixed and certain mind, one confident of belief, will, and power. But under those circumstances, restraint seems an odd enterprise, like rushing outside to crush a bee seen through a window.

Here reality becomes more complex than Holmes' statement supposes. It is perhaps less the certainty of our convictions that prompts us into intolerance than their uncertainty. Are there really any beliefs or values to which we regularly profess allegiance that we can honestly say we are completely confident of—either of our actual belief in them or of our courage to act on them should they be challenged?[42] Even if we are confident of our own minds, we can rarely be sure of the beliefs of others, on whom in nearly every instance our own beliefs depend for success. The upshot would seem to be that beliefs and will are dynamic, always in flux, inevitably uncertain, and, at least at times, seemingly very fragile. That is why there is such a great need for individuals and communities to engage in expression, to reaffirm beliefs; that is why we encounter an endless fascination in societies with examining just how one would behave under circumstances where belief and commitment to belief (which perhaps are the same thing) are put to the test; and that is why, in the end, people draw the inferences they do about doubt and weakness of resolve from the posture of passivity or tolerance—not because they always coincide but because they do so with enough frequency that it provides a reservoir of human experience from which to draw such inferences.

The trouble with speech behavior, therefore, is that it very often demands a response from those who know of it. It compels us to act in response, and in that sense it exerts a controlling power over other people's behavior. It is agenda-setting, for without any response, messages different from those we want to be communicated are communicated. We can no more afford to ignore these feelings than we can the need to mourn the death of someone close. Thus, not only can speech lead us into errors of judgment by playing on our passions, as we recognize in the context of trial, and in that way distort the truth-seeking process;[43] it can also distract us from dealing with other issues that ought to have a higher priority or, for that matter, distract us from pursuing other goals we may seek alongside the search for truth (social harmony, collegiality, mutual respect, and so on).

The entire subject of interaction between speakers and listeners is, therefore, a matter of considerable complexity. We trivialize the problems speech behavior can pose for any individual or community by speaking simply of the risk of the speaker's persuading some weaker minded listeners or of the offense that some listeners will experience at the speech, as something equivalent to a momentary pain one feels from a cut or a stubbed toe. Pain is experienced when others speak in ways we feel are wrong, as genuine and as real as the pleasure we feel in others' happiness or the sadness in their misfortune, arising perhaps from the same emotional capacity of imaginative projection, or what we call empathy. But it is also the case that the identities of those witness to an event become implicated, and to some extent controlled, by the event itself. The values and beliefs that together define the group or broader society exist in a state of some flux, in which both belief and the willingness to act on them and in their defense must be settled and resettled again and again.

In this general way, therefore, speech can produce important harms. This, of course, is clearly revealed when speech is explicitly insulting or threatening to particular individuals. The identity of these individuals is now in jeopardy not only because others may be "persuaded" to think less of them but also because others will be watching to see how they respond. The problem is hardly insignificant. However much as individuals we may try to disconnect our own feelings about ourselves from the feelings that others bear toward us, we are never more than partially successful, as Isaiah Berlin noted in a well-known essay:

> For am I not what I am, to some degree, in virtue of what others think and feel me to be? When I ask myself what I am, and answer: an Englishman, a Chinese, a merchant, a man of no importance, a millionaire, a convict—I find upon analysis that to possess these attributes entails being recognized as belonging to a particular group or class by other persons in my society, and that this recognition is part of the meaning of most of the terms that denote some of my most personal and permanent characteristics. I am not disembodied reason. Nor am I Robinson Crusoe, alone upon his island. It is not only that my material life depends upon interaction with other men, or that I am what I am as a result of social forces, but that some, perhaps all, of my ideas about myself, in particular my sense of my own moral and social identity, are intelligible only

> in terms of the social network in which I am (the metaphor must not be pressed too far) an element.[44]

The interconnections between the individual self and social perceptions is true for groups of people as well as for individuals, which, of course, is why racial and religious slurs are so hurtful.

In fact, we are routinely concerned, and properly so, with the way other people are thinking about us, whether it be revealed by speech or other behavior. For confirmation of the reality of this kind of injury we need look no further than to the jurisprudence of free speech itself, to what laws have been sanctioned, and even to the types of arguments that are made in support of free speech positions. For example, our rules against defamation and the publication of embarrassing personal facts, as well as to some extent our laws of private property, which help secure information against public disclosure, recognize the fact of the human condition that one's self-esteem is to a considerable degree related to what those around one think. Even more important, what people think is in some important measure dependent on one's response, so that to control the response is to control what others think. This fact has sometimes even become the basis for free speech claims, as individuals or organizations compelled by the state to give assistance to speech they disapprove of have argued that their own free speech rights have been thereby infringed. [45]

Even free speech theory itself makes rhetorical use of our fundamental concern with the minds of others—though we seem unable to recognize it as such because of our tendency to give excessive weight in developing free speech theory to our interest in obtaining information. It is sometimes said, for example, in defense of the free speech principle that a government's decision to censor information because of a distrust of the citizens' ability to use the information properly interferes with our interest in maintaining our personal "autonomy," which is defined as making our own decisions (including whether to obey the law) after receiving all available information. Such a view has powerful intuitive appeal. Its appeal rests not only in the description of us as being interested in obtaining all the information we need to make good decisions. It is partly that but not entirely, for it is clear once we examine our own lives that we do not always sacrifice other interests we have in order to acquire all available information.[46] The appeal of the argument is rather in the need we feel to object to, and forbid, the government's act because it is based on a view of us as incompetent. The thought is insulting and demeaning. We may not care at all about having the particular information that would have been communicated by the censored speech, but we care very much that the government not think of us as incompetent, just as we might very well feel insulted at not being invited to a dinner party we would actually dread attending. Our own sense of ourselves, our identity, is at stake; and we also think that if officials are taking this view of us, then we had better stop them now before they act on it in other ways too.

Thus, free speech argumentation itself sometimes makes use of our natural and proper concern with the thinking at work in other people. We should also learn from this the importance of not confusing the real basis of our objection with other, more immediately practical interests, such as our interest in acquiring information. Were we to decide ourselves what to hear and what not to hear, we might very well choose not to listen to the very same speech the earlier censorship of which we so vigorously objected to; we might decide to forbid it for precisely the same reason, namely, that the speech act was insulting and threatening.

These general observations about the injury speech is capable of inflicting can be tested in the concrete context of *Skokie*. We have already noted how speech may cause us to fear for our own safety and well-being. The fears of anti-Semitism being stimulated by the willingness of a group to publicly don the Nazi uniform and then parade in a heavily Jewish community cannot be dismissed as illogical. Indeed, with some speech acts we easily grasp this psychological reality and, on that basis, forbid such acts by law. During the *Skokie* litigation, for example, many Jewish residents of the community received anonymous telephone calls in which the caller would issue threats and anti-Semitic statements. We can all appreciate the fear these people must have felt for their own personal safety and for those near to them. The injury is similar to that suffered when a person is the victim of a crime, as when a burglar enters the home in the night. The property loss experienced will often be minimal when compared with the emotional, or psychological, injury inflicted by the threat implicit in the burglar's act. In all likelihood, the victims will long thereafter suffer the fears and pains of this personal violation, which consists in part of the greatly enhanced sense of their own personal vulnerability to a repetition of this crime and others they will now so vividly imagine and in part to internal doubts about how they should or would respond to those violations.[47]

The costs of these acts, however, are social as well as personal to the specific victims, and this is where the interactive, or dynamic, character of the speech acts and the identity of the community come into the foreground. For it is the community's response (or nonresponse) to the harmful speech act that will in some measure define the community—a reality that sociologists, beginning with Emile Durkheim, have long observed in the context of the criminal law system (even going so far at times as to suggest that communities actually use criminal behavior as an opportunity for self-definition).[48] The belief in the wrongness of the acts and the will to protect the individuals and the society against future harmful acts may be thrown into doubt, and like it or not, the issue must and will be addressed. Unless the doubts are resolved, the very anticipation of dissolution can alter the conditions of the community itself, affecting it in countless, untraceable ways—as we periodically see with rises in the popular fear of crime, which can alter the political landscape through changes in voting patterns (a phenomenon that German Nazis deliberately utilized in their drive for power in the 1920s and 1930s).[49]

On a small scale this was dramatically brought to the surface in the *Skokie* litigation—though it was surprisingly mishandled by the various participants, thus further indicating the difficulty we have in integrating into the classical analysis of free speech issues a sophisticated understanding of the potential harm from speech. One of the arguments advanced by the city was that it should be entitled to prohibit the march because it would disrupt the community's fair housing program.[50] The claim never got anywhere, ultimately slipping by the wayside as if it were something of a makeweight—which, on the face of it, it certainly seemed to be, since it was hardly self-evident how the march would seriously interfere with a community policy of creating a housing market free of racial prejudice. Yet, as one reads through the record in the case, and particularly the testimony offered by the mayor of Skokie, one can see that a perfectly sensible—really quite powerful—argument was contained within this superficially slight claim. Regardless of whether the argument was ultimately true or not in this case, it was certainly plausible and sufficiently important to warrant careful attention.

In essence, the city's claim was this: The population of Skokie was more or less evenly divided between Jews and Christians, two groups that had tried over the years to live together harmoniously. The relationship, however, had always been fragile, sometimes only a matter of simple coexistence, sometimes almost erupting in hostilities. The city had also tried to accommodate a growing black population within its boundaries, as the migration of blacks from the inner city of Chicago had steadily brought them to Skokie and an adjacent suburb to the north. But these efforts at maintaining a peaceful community environment between potentially hostile groups were unfinished by the time of *Skokie*; tensions remained beneath the surface. What the city feared, therefore, was that the Nazi march would activate these other tensions within the community.[51]

Why there should be a link between the Nazi march and these other tensions may be difficult at first to see. But a potential link is certainly there. The events in *Skokie* raised a question about the prevalence of anti-Semitism generally in the society, not only in its extreme form as represented by Collin's group but in its less overt forms as well. Questions were always hovering about the case: about the depth of our sensitivity to the interests of those who would be most directly injured by the speech (primarily the concentration camp survivors in Skokie), about our rejection of this kind of racial hatred, and about our preparedness to act against more explicit manifestations of it. There was, conceivably, even a question in the minds of some Jews that some people perhaps actually wanted them to suffer the injury of the Nazi march—a motivation that need not have been anti-Semitic at root but could have arisen out of anger over issues having little or nothing to do with racial or religious prejudice.

The issues raised by *Skokie*, therefore, transcended the simple possibility that the Nazi march would persuade some members of the audience or the general public to embrace all or part of the Nazi doctrine or that it would offend its immediate audience. The event bore the potential for enormous symbolic meaning, whatever was done by the various participants in the drama. (The possible connections between the particular events in Skokie and the general level of anti-Semitic behavior in the larger society were, in fact, noted in the popular press at the time. There were, for example, reports of an increase in attacks on synagogues and members.)[52]

Thus, Wigmore was certainly correct when he pointed out that it is inappropriate to deny society the right to prohibit and punish certain speech simply because that particular speech act was unlikely to be successful. A technical violation can encourage others to act similarly, as Wigmore said, but there may be other reasons why we might choose to ignore the likelihood of success in administering our criminal laws, as in fact we routinely do. Sometimes, when viewed in isolation, events can seem insignificant and unworthy of people's attention, but actually, for whatever combination of reasons, they have been invested by the society with great symbolic significance. What has happened before, or contemporaneously with, the particular act—the overall context in which it occurs—will define its social meaning. Events frequently take on such extra baggage of meaning for people, and when this occurs, it is unfair to segregate them from their context and denude them of the real meaning they possess.

We must appreciate the depth and the power of the conflicts people face when confronted with behavior they regard as wrong, immoral, or injurious to others. For each individual as well as the community, the response taken is self-defining. By a tolerant or divided response, in particular, the individual and the community can be made to feel implicated in, and their identity tarnished by, those acts. A subtle but nonetheless extensive process of dissolution can then occur by denying the individual or group the means of responding to the behavior they find troublesome. People must have some means of demonstrating commitment to their beliefs and determination to defend them. We recognize and act on these needs all the time, especially as we set about the business of enacting laws and punishing violations. They are what induced President Jimmy Carter to boycott the 1982 Olympic Games, a policy borne (at least in part) of a feeling that to pursue a business-as-usual course in the face of the invasion of Afghanistan would implicitly sanction that act of aggression. They are what led the justices of the Supreme Court in *Brown v. Board of Education*, and its sequels, to want to present those decisions as the results of a unanimous and undivided bench. They are what made the judges in the *Skokie* case announce their own "personal" views about the speech they found it their professional responsibility to protect.

But there is another objection to punishing speech acts to be considered. Assuming we agree—it is said—with the arguments just advanced about the complexity of the harm from enforced toleration of speech acts, it remains the case that you possess other means, besides legal prohibition, by which to satisfy the needs of intolerance. Most important, you retain the opportunity to express your own positions on the issues raised. So long as that opportunity is available to you, you need not possess the power to silence others from speaking as they wish.

This is a commonly heard argument in free speech disputes. At the outset we should note that the availability, by itself, of other means of accomplishing the ends we seek does not explain why we may justly be deprived of some. But it is also doubtful that any real equivalence exists between official restraints on speech and other means of self-help. We suffer a serious loss when we strip ourselves of the use of legal restraints against speech behavior we regard as socially destructive. It is easier to organize an official response through the lawmaking process. Procedures for this are established and readily available. But its primary value for us is as a *communicative* tool for satisfying the need of the community to express its position on the issues raised by the speaker and in doing so to define and create itself.

Law plays a special role in this country. It provides a process through which we create a social identity, by which we reflect and embody the aspirations and values of the community. Enforcement of the law bespeaks a commitment to those aspirations and values.[53] We all know the importance of maintaining through our public institutions a level of behavior that rises significantly above what we otherwise insist on. That is why public officials must resign when they say things in public that demonstrate an insensitivity to the values we are trying through our public institutions to achieve. To some degree, to be sure, the public posture the society presents to the world exists in many layers and not all of them are official. Other individuals, groups, and institutions within the society—such as religious organizations and private societies—sometime speak "for" the general society in the way that law does. It is true, for example, that in *Skokie* such groups did come forward to address the issues involved, and in doing so helped to assuage the fears of many people on the matter of anti-Semitism. Still, it seems reasonable to think that we gain by keeping open to ourselves the possibility of responding to behavior we believe harmful through our lawmaking process.

## IV

To conclude that speech activity can cause very serious social harm is not to say that all speech that causes harm ought to be prohibited. The same point works the other way, too: The fact that speech can bring valuable benefits to us, as individuals and as a community, does not mean that we ought to restrain ourselves from ever regulating it. The problem is finding the right balance, which, of course, requires that we make a judgment. We balance the harms against the benefits with respect to nonspeech behavior, and we make judgments about what should be free and what restricted. If we were to do the same with speech activity—to feel free to render a true accounting of the benefits and harms—would we arrive at the results we now do under the free speech clause of the First Amendment? Would we say that the benefits of speech identified under the classical model are so present in the "speech" acts at issue in *Skokie* that they outweigh the full range of injuries likely to be experienced by those acts?

At the very least, it seems unsatisfactory to construct our thinking about free speech so heavily, and so uncritically, out of the building blocks excavated from another era, when the needs of the society and the materials available were potentially so different from those in our own time. However useful it may be as a heuristic device to take the traditional New England town meeting as a paradigm for thinking about free speech in twentieth-century United States, as Meiklejohn did, there are serious risks of oversimplification involved in making that our comparative model. Close, tightly knit communities possess a degree of control over their members that is simply lost in a society numbering in the hundreds of millions of people. Homogeneity is no longer the central characteristic defining its people, but rather its opposite.

At times with the conventional libertarian defense of free speech, there can be a virtual denial of any social risk like propaganda and manipulation of public opinion by speech, despite the repeated experiences of other countries in this century to the contrary. Indeed, the Pollyannaish claim that the truth will always win out as a natural result of evolutionary processes deserves the brushing aside that Alexander Bickel gave it when he remarked: "[W]e have lived through too much to believe it."[54] Only by recurring to the empty plea that we think in terms of the "long run" can anything be salvaged from the argument, and to nearly all of us, it is rightly said, the long run will always come too late.

There have been few serious attempts to integrate into the general free speech discourse a more complex and realistic view of modern society. After World War II a few writers discussed the need for a reappraisal of the classical assumptions about the impact of propaganda and manipulative political rhetoric on political behavior, noting how the development of a mass communications media and the rise of a new "mobile public opinion" made a wide-open attitude toward speech a more socially precarious policy.[55] The Supreme Court's decision in the *Beauharnais* case,[56] mentioned in chapter 1, seemed to take some stock of this way of looking at things, as did a few other opinions. Such a perspective, however, has never really been woven successfully into the fabric of our free speech jurisprudence. Today one still finds the typical encomiums to the capacities of citizens to handle their own affairs, with the asserted role of the courts being simply to facilitate that process by keeping the government at bay.

The consequences of this simple approach to free speech questions in the twentieth century cause us to stumble on both sides of the problem, in appraising both the potential benefits we derive from protecting speech and the potential harms. If we conceive of the benefits as those we derive from the best uses of speech, it is not surprising that we have little to say at the margins, where these values will necessarily be attenuated and flooded with other concerns. The merits claimed for free speech are going to have difficulty bearing the full weight of the harms that can be expected. Not surprisingly, a tendency is exhibited at this point to understate those harms.

This kind of refusal to think openly and critically about problems connected with the idea of free speech is present with respect to the other two fundamental issues for free speech theory noted at the end of chapter 1—namely, the problem of justifying the special protections afforded speech and not other, nonspeech acts[57] and the problem of justifying the placement of the interpretative and enforcement functions connected with a free speech principle in the hands of the judiciary. We shall defer addressing these problems for the moment, keeping our attention fixed for the time being on the issue of the degree to which we should limit our authority to regulate speech activities.

Free speech thought is possessed of a kind of schizophrenia, however, and there is another side to it that we frequently encounter in the theoretical accounts and in the cases. It differs rather fundamentally with the premises of the classical model, at least on certain basic issues. It is a vision of free speech, however, that works largely out of public sight, a kind of north slope of First Amendment thought. Here, where the sun's rays penetrate only indirectly, we encounter a more complex and less naive understanding of the role of free speech in modern society. With the more official rhetoric and vision of the classical model, it is regularly engaged in hidden tension.

# 3

# The Fortress Model and Its Limits

Even though in any particular case the positive benefits from having certain speech may not exceed the injury that protection of it will bring, there may be other reasons for protecting that speech—strategic reasons. This is at least partly what we mean when we say in free speech disputes that "a principle is at stake"; to secure that principle we must be prepared to include within its compass some unworthy speech.

In chapter 1 we considered the possibility that one could justify protecting extremist speech on the ground that the administration of the free speech principle required a fixed rule both for convenience and for ensuring against inadvertent incursions into the terrain of valued expression. In chapter 2 we concluded that a commitment to self-government entailed a decision on our part not to let the government usurp our power of decision making over the proper limits of speech activity in the society; and it entailed a concomitant concession by us that we should be restricted in our own power to *hand back* to the government the power to restrict speech activity minimally necessary to meaningful exercise of our self-governing role. Suppose, however, the problem is not simply one of administrative convenience or insurance against mishaps or appropriate allocation of authority, but rather the presence of some power that stands as a constant threat to the basic principle itself and can be expected to take advantage of any opportunity left open to it to push for suppression of truly valuable speech. In such a case, it would seem sensible to create a system that foreclosed any such opportunities. One obvious way of doing that might be to secure the boundary of protected speech at some considerable distance from the speech activity we truly prize.

This way of thinking, in fact, informs much of our contemporary arguments for free speech protection in the extreme cases. We erect a fortress of legal doctrine under which choices over speech regulation are sharply constricted, and we staff the fortress with officials from another governmental branch, the judiciary. But if the purpose of protecting extremist speech under this fortress model is to achieve protection for the inner core, against whom is the fortress needed? Can we be sure that the fortress will withstand that threat? What are the consequences to the larger community of erecting that fortress?

## I

In chapter 2 we saw that a central premise of the classical model is that the government is possessed of only limited powers, which are derived ultimately from the people, who in turn constitute the true sovereign. But a common corollary to this conception of the state is the belief that the government also constitutes an unvarying threat to the liberties of the citizens, including the liberty of speech. By this vision, the government stands in a perpetually antagonistic and hostile position toward the larger society. Power corrupts, the saying goes. Every government bears within its personality an atavistic longing to recapture the autocratic powers of its ancestors. This reality, it is said, must be included in our calculations about how to structure something as important and as vulnerable as the First Amendment.[1]

This asserted reality of an antagonistic government affects the contours of the principle of free speech, playing an important role in determining its scope and the strategy behind its implementation.[2] It also provides a justification for judicial enforcement of the principle, since it is desirable in the context of this ever-present external threat to have an institution within the governmental structure itself, with all the attendant power of a state institution, charged with the task of guarding against excesses by the other two branches.

The dangers that must be guarded against are subtle as well as obvious. If one believes the government will be tempted to take advantage of whatever opportunities are available to it to silence what it dislikes, and especially that which is critical of it, the problem cannot be disposed of simply by standing guard over the core area of valued expression. This is so for three primary reasons. First, much damage can be done before the unconstitutional behavior can be corrected. Lawsuits take time and money to defend; they are slow, and by the time judges act, the importance of the speech for the individual or for the society may have receded. Second, the law must take account of its own limitations, and one of its most important is its limited capacity to ascertain the truth in any factual dispute. The judicial system may make mistakes, especially as it seeks to reconstruct a past reality.[3] Speech activity, moreover, is easily intimidated by the prospect of coercion. In recognition of all this, it is naturally viewed as advisable to adopt a rule under which speech is always given the benefit of the doubt. Third, again taking stock of the legal system's own limitations, we must realize that judges, being human, will not only make mistakes but will sometimes succumb to the pressures exerted by the government to allow restraints that ought not to be allowed. To guard against these possibilities we must give judges as little room to maneuver as possible and, again, extend the boundary of the realm of protected speech into the hinterlands of speech in order to minimize the potential harm from judicial miscalculation and misdeeds.

Without these internal protections the law itself may fail to perform its fundamental assignment of protecting the liberty of speech. The very likely prospect of failure would itself have an inhibiting effect on expression, for, given human nature, many people will choose not to exercise their liberty for fear they would be the victims of those systemic failures in an environment in which the state is so inherently antagonistic to the exercise of liberty. "A rule compelling the critic of official conduct to guarantee the truth of all his factual assertions," reasoned the Court in *New York Times Co. v. Sullivan* about the prospect of libel judgments, "leads to a comparable 'self-censorship.' "[4]

## II

There is certainly some truth in this common argument about the need to take cognizance of the threat of government abuse and of the inhibiting effects on discussion of legal rules that are too closely tailored to the benefits and harms of individual cases. But it is an argument that is seriously overplayed in twentieth century life. As it is generally put, the problem is with a government's departing from and turning against its democratic origins. The seriousness with which this notion is held, however, and the earnestness with which it is advanced as a basis for structuring the First Amendment seem out of proportion to its reality as an actual problem. To be sure, we have instances in this century, and within the last decade, that give the idea more than hypothetical credibility. Perhaps the Pentagon Papers case[5] may reasonably be regarded as an instance in which the government acted to thwart, rather than to advance, democratic interests. Such examples pale beside the number of instances in which suppression of speech has been not in contravention to majority wishes but in furtherance of them—a point emphasized in chapter 2. To say that, however, is to raise another possibility: that in constructing a social system in which valuable speech activity is secure, we must take account of a reality in which both the government and the people themselves are threats to that goal. If there is a problem of a tendency to excessive intolerance toward speech activity, it would seem to be not with "the government" alone but with "the people" as well, acting through their government.

The record of periods of excessive repression against speech supports this judgment. Where the regulation of speech has gone awry, the restraints have arisen primarily through the system of majority rule. Chafee reached this conclusion in his famous study of excessive intolerance in the years from World War I to World War II. It was the people who prodded the government into action and not the government acting independently and on its own motion.[6] A similar phenomenon occurred during the McCarthy years of the early 1950s.[7] And the tensions of the 1960s and the Vietnam era were between groups of citizens, not just, or even primarily, between the citizens and the government.[8]

But to note and take account of the phenomenon of public intolerance is to shift significantly the terms of our thinking about free speech. A new perspective emerges, one that retains a principal premise of the classical model—that the central object of free speech is the protection of valuable speech, defined primarily in terms of the communication of information and ideas in the process of seeking truth—but departs from the premise that the people themselves are trustworthy, at least insofar as decisions over the regulation of speech go. This shift in our thinking has radiating consequences for all aspects of the free speech principle, for the methods selected for enforcement as well as for the scope of protection sought.

The real threat to liberty of speech, then, rests within the general population of citizens instead of with officialdom alone. In fact, this has been a basic fear of democracy from the beginning. It arose in the debates and discussions of the Framers of the Constitution and was reflected in the structure chosen for the government to be formed in America.[9] It is a recurrent theme in the writings about political theory of the nineteenth century. Authors like Alexis de Tocqueville and Mill wrote of how the incoming tide of democracy had washed up the problem of the "tyranny of the majority," which could be every bit as threatening to basic human liberties—even more so—as the despotic political regimes that democracy had replaced.[10]

For a while, Mill said, people had assumed that tyranny was at an end, because democracy brought an identity of interests between the acts of the state and the will of the people. But success can "disclose faults and infirmities which failure might have concealed from observation."[11] As democracy flourished, it brought with it the realization that "such phrases as 'self-government,' and 'the power of the people' over themselves, do not express the true state of affairs. The 'people' who exercise the power are not always the same people with those over whom it is exercised; and the 'self-government' spoken of is not the government of each by himself, but of each by all the rest."[12] Since the majority of the moment "may desire to oppress a part of their number, . . . precautions are as much needed against this as against any other abuse of power."[13]

Just as to Mill intolerance was "so natural to mankind," so it was to the English political economist Walter Bagehot, whose 1874 essay on toleration remains influential today. Bagehot praised England for having increased its toleration for the "expression of opinion," yet he still felt the need to write an essay defending toleration because a tendency to relapse into persecution was integral to the human personality:

> What was said long ago of slavery seems to be equally true of persecution—it "exists by the law of nature." It is so congenial to human nature, that it has arisen everywhere in past times, as history shows; that the cessation of it is a matter of recent times in England; that even now, taking the world as a whole, the practice and the theory of it are in a triumphant majority. Most men have always much preferred persecution, and do so still; and it is therefore only natural that it should continually reappear in discussion and argument.[14]

These sentiments have been repeated in the twentieth century, though generally they take a somewhat less explicit, more submerged form. We saw earlier that Holmes found intolerance "perfectly logical." In fact, some of his private correspondence at the time, primarily with Learned Hand, gives fuller statement to this perspective on the naturalness of intolerance.[15] Hand himself, speaking of the unfavorable conditions of the time for free speech, deplored the "existing hysteria" and the "merry sport of Red-baiting" and noted how "the pack gives tongue more and more shrilly."[16] The public origins of the intolerance of that period, which Chafee later documented, did not escape the attention of either Hand or Holmes at the time.

Meiklejohn, too, was painfully conscious of the reality of public intolerance. While at times he adopted the customary oversimplified vision of free speech as simply protecting the self-governing process from "government" interference, he was plainly aware of the more complex origins of the censorship he deplored. At the beginning of his essay, Meiklejohn refers to the rising level of intolerance across the country, with roots sunk deep in general popular sentiment. The government activities he objected to—the FBI investigations, the legislative inquiries into "un-American" activities, and the like—were, in fact, supported and encouraged by a sympathetic public. Of the FBI "system of espionage," Meiklejohn wrote of how "that procedure reveals an attitude toward freedom of speech which is widely held in the United States. Many of us are now convinced that, under the Constitution, the government is justified in bringing pressure to bear against the holding or expressing of beliefs which are labeled 'dangerous.' Congress, we think, may rightly abridge the freedom of such beliefs." To Meiklejohn these were the "wretched days of postwar and, it may be, of prewar, hysterical brutality, when we Americans, from the president down, are seeking to thrust back Communist belief by jailing its advocates, by debarring them from office, by expelling them from the country, by hating them."[17]

The phenomenon of public intolerance, of the potential tyranny of the majority, then, can be seen weaving its way throughout the thought about freedom of speech. One might even say that no more disparaging, and despairing, view of the nature of the average person, of his and her natural tendency to be intolerant, is to be found than in the libertarian literature on the subject of free speech. The impulse may ebb and flow, possibly affected in its movements by the moons of economics or of war, but it is ever-present and even ineradicable, so deep is it etched into the human character.[18]

In the libertarian literature, the nature of this impulse to intolerance is usually said or thought to rest in the deeper tendency of people to need to believe and to have others believe similarly. "The likings and dislikings of society, or of some powerful portion of it," Mill wrote, "are thus the main thing which has practically determined the rules laid down for general observance, under the penalties of law or opinion."[19] "All silencing of discussion is an assumption of infallibility,"[20] he added with unfortunate overstatement. Bagehot argued similarly: "Persons of strong opinions wish, above other things, to propagate those opinions. They find close at hand what seems an immense engine for that propagation; they find the *State*, which has had a great and undeniable influence in helping some and hindering others—and in their eagerness they can hardly understand why they should not make use of this great engine to crush the errors which they hate, and to replace them with the tenets they approve."[21]

A common accompaniment to this association of repression with belief is the notion that beliefs are simply the product of circumstances in which one was raised as a child and lives as an adult—the social and intellectual milieu of whatever one's existence happens to be. This view of the social conditioning of belief is widely reflected in the general literature. When Socrates opens his defense, in the *Apology*, against the charge that he has corrupted the Athenian youth, he warns the jurors of their inclination to prejudge him. Their attitudes toward him, says Socrates, were instilled in them while they were just children—when these "dangerous accusers" had "approached . . . [them] at the most impressionable age, when . . . [they] were children or adolescents, and they literally won their case by default, because there was no one to defend me."[22]

Mark Twain called these kinds of beliefs "corn-pone opinions": those opinions people hold, not because they have reasoned them out for themselves or because they are derived from firsthand experience, but because a person "must think and feel with the bulk of his neighbors, or suffer damage in his social standing and in his business properties." Perhaps the same roots in the human psyche that make our self-esteem a product of what others think of us—and, as we noted in chapter 2, provide the primary justification for libel and invasion of privacy laws—also create the need to believe generally as others do. In any event, to Twain corn-pone opinions derive from the "inborn requirement of self-approval," which, "as a rule . . . has its source in but one place and not elsewhere—the approval of other people." It is the aggregate of corn-pone opinions, he wrote, that together make up "Public Opinion," which is "held in reverence" and which "[s]ome think . . . the voice of God."[23]

Mill offered a less amusing, though no less severe, intellectual indictment, finding that most people define their system of belief according to that held by their "world," which "to each individual means the part of it with which he comes in contact: his party, his sect, his church, his class of society; the man may be called, by comparison, almost liberal and large-minded to whom it means anything so comprehensive as his own country of his own age."[24] To such people beliefs are held with unshakable conviction of their rightness: "He develops upon his own world the responsibility of being in the right against the dissentient worlds of other people and it never troubles him that mere accident has decided which of these numerous worlds is the object of his reliance, and that the same causes which make him a churchman in London would have made him a Buddhist or a Confucian in Peking."[25]

Bagehot found other, deeper fears behind the need and desire to persecute others for their beliefs and for what they say. The "main motive behind persecution," he argued, was "a most ancient political idea which once ruled the world, and of which deep vestiges are still to be traced on many sides."[26] This is the fear, most profoundly expressed in societies with a strong religious base, that "any member may by his acts bring down the wrath of the gods on the other members, and, so to speak, on the whole company."[27] There is a profound fear of misplaced attribution of responsibility for the sins of others: "And as false opinions about the gods have almost always been thought to be peculiarly odious to them, the misbeliever, the 'miscreant,' has almost always thought to be likely not only to impair hereafter the salvation of himself and others in a future world, but also to bring on his neighbours and his nation grievous calamities immediately in this. He has been persecuted to stop political danger more than to arrest intellectual error."[28]

Like Mill, however, Bagehot also saw in modern societies an extensive process directed at the inculcation of belief, one which bore the bitter fruit of excessive intolerance. He emphasized a relationship—a view that would be restated again in the twentieth-century literature on child psychology[29]—between childhood and adult intolerance: Societies require a "common religion," he argued, in the secular as well as religious sense. Laws take on the character of "holy customs," which are essential "to aid in creating a common national character, which in aftertimes may be tame enough to bear discussion, and which may suggest common axioms upon which discussion can be founded."[30] So it is that "[e]very parent wisely teaches his child his own creed, and til the child has attained a certain age, it is better that he should not hear too much of any other."[31] This elemental need for accepted beliefs is both psychologically and intellectually desirable, for the child "will hardly comprehend any creed unless he has been taught some creed."[32] But the result is that people arrive at the "age of discussion" not with a tabula rasa but with an elaborate set of instilled beliefs.

The view of an impulse to intolerance related to a need to believe is repeated in some of the twentieth-century literature referring to free speech. There is recognition of a powerful desire among people, among societies, to insist on conformity of opinion. In his writings, both official and personal, Holmes continually detects this tendency in himself and struggles against the instinct to believe whatever it is he happened to come to believe; sometimes the impulse is overpowering and he is forced to concede defeat, referring to such stubborn beliefs as his "Cant Helps."[33] But we also know from his famous account of the nature of intolerance in the *Abrams* dissent, which we considered at some length in chapter 2, that the desire or need to establish conformity of opinion had a much more complicated foundation than simply the need to believe, and there is some resonance with Bagehot's idea about intolerance arising from fear of doubt as much as from certainty of belief (a theme we shall return to in chapter 5).

We need not pursue this any longer at the moment; the essential point we are looking for has already been made—namely, that in thinking about freedom of speech we cannot ignore the fact that much of human society has yet to accept the principle fully or to make it a cornerstone of their conduct. At work in the human personality is a contrary impulse that makes free speech protection always a very fragile enterprise. Thus, free speech must fight against something deep in human nature that denies the importance of its purposes, whether they be the attainment of truth, the process of self-government and majority rule, or the importance of speech to individual autonomy and self-fulfillment. The "government" provides the means by which this impulse is executed, but the impulse rests elsewhere in the recesses of human nature.

## III

This working notion about the human personality and its relation to the control of speech activity obviously brings an important new dimension to the discussion about the limits of free speech. In chapter 2, we approached the question of defining the scope of the free speech principle by asking how we would assess the benefits and harms of protection in particular cases. A basic premise of that discussion was that we all share a capacity for reasonable thought about that issue. In that context I argued that in arriving at our final assessment, it was necessary to take into account the importance to us of maintaining a social sense of shared beliefs and values and of using the lawmaking process for that function. Now, however, that point has been turned back against us. It is said that that very need of people has a tendency to get out of hand, to become excessive, and when it does, we must expect serious encroachment on the treasured freedom of open discussion. It is a question of degree. "We" (that is to say, some of us) might be able to strike a good and proper balance between the potential benefits of speech and the maintenance of other basic values, recognizing that while mistakes would occur, they would probably be of only marginal significance. That is not something we can leave to the general society, where the impulse to intolerance runs too deep and, if given too free a reign, will threaten to disrupt any reasonable balance struck between free speech and other interests.

The presence of this distorting characteristic in the human judgment about speech activity exerts a powerful and decisive effect on our minds as to how we talk and think about the principle of free speech. As with the perceived perpetual threat of a tyrannical government, the asserted reality of excessive public intolerance introduces the element of strategy into the design and implementation of the free speech idea. The strategy now, however, goes beyond what we saw before. Protection of extremes of speech still provides a kind of buffer zone, though now against a more broadly based threat. This speech is protected not for its intrinsic value but because that protection better insulates what is intrinsically valuable. Along with this, however, comes a whole panoply of doctrines and rhetoric.

The threat of public intolerance is the basis for treating free speech as a "principle" and, further, for analyzing speech cases in terms of subject matter "categories"—political, commercial, and the like. The method of reasoning through a particular dispute then involves deciding whether the speech is on one subject or another, rather than deciding to what extent the speech yields incremental gains to the purposes of free speech and weighing that against the competing social harms caused by the protection of that speech. Judges are instructed to be blind to the "worth," or "value," of the speech involved in any individual case, an inquiry said to be beyond their competence to resolve and, in any event, illegitimate to entertain.[34] This approach helps explain why Justice Lewis Powell could write in a well-known libel decision that "[u]nder the First Amendment there is no such thing as a false idea,"[35] and why the judges in *Skokie* could, after professing their own belief in the utter worthlessness of the speech they were about to protect, go on to discount as irrelevant their own personal views, because "[i]deological tyranny, no matter how worthy its motivation, is forbidden as much to appointed judges as to elected legislators."[36] To adopt this approach, it is argued, reduces the opportunities for litigation (which, as noted earlier, can itself be detrimental to speech interests, quickens the speed with which litigation begun may be halted), and lessens the risks of inadvertent or deliberate misweighing of free speech and social interests by judges under pressure from public intolerance.[37]

But the impact of this perceived threat of public intolerance on the shape of the free speech principle transcends the doctrinal and, in essence, transfigures its very presentation to the world. It explains some of the otherwise inexplicable, sometimes even bizarre, aspects of our discourse about free speech. Much of what is said about free speech is designed—sometimes skillfully, sometimes not—to induce people to think about the principle in ways that will limit or impede their inclination to abandon it in favor of the intolerance impulse. Thus we have the legerdemain of those, like Justices Hugo Black and William Douglas, who assert with so much confidence and even fervor that the language of the First Amendment makes it an "absolute" prohibition on government interference, admitting of no exceptions.[38] We have unfounded assertions about the original, historical meaning of the free speech idea, when there is little or no supporting evidence for such assertions, and in some instances even contrary evidence.[39] We have the equally unsupportable claims about how prior cases foreclose consideration of any claim to regulate speech. And we have the attempt to create an image of the government as a constant threat to all our liberties when the reality is less than the appearance created.[40] Such exaggerations and overstatements arise, at least in part, from the wish to create the appearance of choicelessness about the scope of free speech.

This basic aim of reinforcing the fortress walls by constructing ways to conceive of free speech also helps explain, even if only partially, the true nature and intensity of certain debates within classical theory, some of which were alluded to in chapter 2. Take, for example, the debate over whether free speech should be conceived of in terms of the advancement of the general good, of the whole society, or in terms of being an "end in itself" or a "right" of the speaker that exists independent of, perhaps even in spite of, the benefits that it yields for the larger community. It is difficult to understand this debate if one thinks about it as a dispute over whether some specific speech activity ought to receive protection or not, since protection of almost any speech could reasonably occur under either theory, so open-ended are the concepts of the "general good" and of "individual rights."[41] It is possible, however, to see a larger meaning to the dispute if one looks not at the handling of specific cases but at trying to induce a way of thinking about free speech. What is objectionable or troublesome about the collective-interest aproach is that it tends to foster a way of thinking about free speech that involves or requires explaining and justifying how speech protection actually benefits the society (which is a most difficult task), that it creates the impression that circumstances may change and so invites continual reexamination of the principle, and that it encourages the view that free speech is only a means to other ends, making it seem therefore more easily dispensable by the simple decision that those ends can better be achieved by other routes.

The foundation of the concern, in other words, is to be found in doubts about the process of rational thought, about a tendency in people to rationalize their irrational wants (or, less deceitfully, to misconceive their own best interests), and finally, perhaps, about the capacity of language and speech to capture and communicate those true interests. Given this view of the world, it seems both sensible and imperative to take a position in defense of free speech in which essentially no thought is permitted: free speech is simply a "right" that each individual possesses against the larger society and that need not be defended on any other basis. It is elemental, beyond argument, a priori.

On the other hand, from the perspective of the collective-good approach to thinking about free speech, there is a different implicit judgment or calculation about the conception that will best forestall abandonment of free speech. What is claimed on behalf of the "rights" approach as an advantage—the elimination of a need for justification—is seen as a disadvantage. For, only as long as people think that *their* own best interests are being advanced by a free speech principle will they agree to adhere to it.[42] Any acknowledgment of overall disadvantage, especially when accompanied by such an open-ended claim of a simple "right" to speech, will only accentuate the difficulties people will have with controlling the intolerance impulse in the first place.

Another manifestation of the impact on free speech rhetoric caused by the concern with public intolerance can be found in the characterizations of the postures of tolerance and intolerance. In the free speech idiom (as we partly noted in chapter 2), intolerance is commonly associated with fear and tolerance with fearlessness. Louis Brandeis's famous concurrence in *Whitney v. California*[43] illustrates this kind of portrayal of the personalities behind the competing positions on protection and suppression: "Those who won our independence by revolution were not cowards. They did not fear political change. . . . Fear of serious injury cannot alone justify suppression of free speech and assembly."[44] Similar associations appear in Meiklejohn's essay: "When a question of policy is 'before the house,' free men choose to meet it not with their eyes shut, but with their eyes open. To be afraid of ideas, any idea, is to be unfit for self-government."[45] Later he writes: "We Americans are not afraid of ideas, of any idea, if only we can have a fair chance to think about it."[46] Both men were employing a style of argument in which the desired behavior is encouraged through the manipulation of images and general characterizations of both the wanted and the unwanted behavior.

As we follow the thread of thought about the risks of public intolerance, then, a new understanding of the structure of free speech, including the role and function of protection of extremist speech, emerges. Is it justified? Is it likely to be effective?

## IV

In assessing the merits of the fortress model, we might begin by considering how it deviates from the classical model and how it might be problematic for us to use it as the foundation for the social institution of law.

I have already suggested that the fortress model tends to play a submerged role in public dialogue about free speech, and if one pauses to consider them, the reasons for this will become quickly apparent. It highlights a crisis of legitimacy for the judicial branch, for the courts are now no longer serving in the comfortable role of implementing the choices of the citizens, as defined by the classical model, or of protecting the people from "the government," as the limited fortress model presumes. The simple agency role in which courts are so inclined to cast themselves in describing their functions under the free speech principle has been severed, and the unelected, life-tenured attribute of judicial power stands out like a theoretical sore thumb. (In an effort to reestablish the "protector" status, courts occasionally will shift to a self-described role of protecting "a minority" against the democratic majority, but that cannot alone amount to an answer to the judicial dilemma, nor is it a posture that is assumed in the typical extremist speech case.)

Beyond the matter of the legitimacy of judicial authority, the fortress model has the additional problem of contradicting one of the central premises of the inherited classical model—that people are rational, capable, and worthy of trust. The fortress model builds upon an opposite vision of people—that they are moved by irrational impulses and are not to be trusted, not, at least, when it comes to deciding what the limits on speech activity within the society should be. The implication of this contradiction of a basic premise of the classical model, however, goes beyond simply having a different starting point. It casts a shadow over the most important claim traditionally advanced in support of the free speech principle: If the people are incapable of properly weighing the relevant considerations in deciding what should be the proper scope of speech activity, why should we place our trust in the open marketplace of ideas about any other issue? The flaw suddenly revealed in the discussion over the limits of free speech has a more universal import for our understanding of human behavior.[47]

It is important to grasp this point fully, for it guides us to other points as well. What the fortress model observes in human behavior—however intangible a quality it may be—is a deep and profound difficulty in controlling a desire to censor or suppress any difference of belief, opinion, or way of thinking. This quality in human nature manifests itself in decisions about how much liberty is to be permitted in the holding and expressing of belief and opinion. But its consequences can hardly be expected to be contained within that sphere of social interaction alone, and of course they are not. One can easily see how the impulse will spill over into the area of nonlegal restraints toward the expression of opinion. Yet this is not really just a "speech problem," even if one includes the massive area of nonlegal restraints; it would appear to be a problem of *intellectual incapacity*, with an impact that accordingly reaches out across the vast expanse of human behavior.

One would expect to see it manifest itself in decisions over what nonspeech behavior in the society should be prohibited and in decisions about what punishment should be inflicted on those who have transgressed those prohibitions. The impulse to intolerance, in other words, can be expected to have an effect on the entire process of social decision making, both with respect to creating the norms for all social behavior (nonspeech as well as speech) and with respect to the process of punishing violations of those norms. The social need of learning to master or control this impulse that the fortress model identifies seems vast and may even overwhelm into insignificance its limited appearance in the area of legal controls on speech activity.

What are the implications for a free speech principle of this more extended significance of the impulse behind intolerant behavior? The question was raised earlier that this might lead us to doubt the wisdom of entrusting decisions other than speech regulation to the general population—a systemic function it is said to be the primary purpose of the principle of free speech to facilitate. Besides such global challenges to a free speech principle, however, the broader role of the intolerance impulse also raises doubts about the wisdom of pressing the free speech principle to the extremes we have. Initially there are three points to be considered.

First, it might be asked whether recognition of the reality of public intolerance should lead us to think of speech not simply as a liberty to be sheltered from the process but rather as itself constituting a potential threat to that process, as potentially accentuating the irrational tendencies that can undermine the process, and so requiring restraint—precisely the approach we see followed with respect to juries in criminal trials. This observation is especially pertinent to cases of extremist speech, where the gain in the advantage of having a buffer that will help ensure the sanctity of the valued expression must be balanced against the risk that the extreme speech may itself prove effective—that is to say, persuasive. To many, this risk was vividly present, for example, in the *Skokie* case.

Second, we might properly ask whether trying to maintain a free speech policy without any exceptions in order to preserve the security of the basic rule offers only the illusion of success. It will only create a kind of Maginot Line of the First Amendment. There will, of necessity, always have to be exceptions of one sort or another, and so the notion of a complete and unbendable line is entirely illusory. In any case, it should be borne in mind that legal rules can always be changed or manipulated. If a government, with majority support behind it, is really bent on persecuting a particular group, will it really find it significantly more difficult to adopt a new rule favorable to that policy than to extend one already in existence that had been created for a narrower purpose? The claim will invariably be the same in either case: an "emergency" or "special circumstances" will be said to justify the "new" state of affairs. If it is argued that the judges will be there ready to strike down a brand new rule, then one might reasonably wonder why they wouldn't also be there to prohibit the unreasonable expansion of a narrow, legitimate exception. Even if it is thought that a narrow exception poses greater risks of expansion than does a policy against any exceptions (which again is, practically speaking, impossible), is it realistic to assume that the judicial system will summon the strength to resist the tide of pressures from a powerful government intent on suppression? The judicial track record of protection of civil liberties during such periods gives little reason for optimism.[48] It is a sobering fact that even the ACLU succumbed to the rabid intolerance of the 1940s and 1950s by purging suspected Communists from its official hierarchy and by assisting the FBI in identifying "subversives."[49]

These concerns about the chances of securing a real "fortress" of legal rules lead to a third and even more fundamental question. Once we have moved beyond the point of seeing the government as the exclusive threat to free speech—for which a strategy of more democracy might be reasonable to pursue—to the point of seeing most of the people as posing an equivalent or even greater threat, and then to the further recognition that what lies behind that threat is something deep in people's intellectual outlook—in their way of thinking—it would seem no longer to make nearly as much sense to pursue a simple policy of securing the right to speak. In that case, having the legal "right" to speak, when there is no one ready to listen, would seem only an empty possession. Then, for all our single-minded haste to give permanence to a legal principle, we may very well have lost in the process an appreciation that the importance to us of the protected right depends on the maintenance of a host of other rights, too.

Are these criticisms wide of the mark or met by other factors? The risk of persuasion of extremist speech must be acknowledged, but perhaps it is believed that the ideas would get out anyway and so suppression would therefore be fruitless. Or it may still be thought that the risk is outweighed by the added security the protection will provide as a buffer. While it may be conceded that exceptions are inevitable in any doctrinal structure, it remains a reasonable working assumption that the fewer exceptions we have, the better off we are. No approach will be completely secure, but a test offering the least opportunity for linguistic maneuverability is still preferable. Even regimes bent on tyranny sometimes exhibit a surprising tendency to want their actions to appear consistent with preexisting legal norms; so, a principle with the fewest exceptions may have a real payoff.

As for what we can expect to gain by such a partial resolution (securing our liberty of speech) of a more global social problem, a number of responses must be considered: Perhaps the objection is itself irrelevant. We are, after all, simply applying the law—in this case the constitutional law of the First Amendment—and if that turns out to be an inadequate solution to a greater social problem, or even an entirely useless liberty, that is not our concern. Moreover, even with this assumed social reality, perhaps there are other positive gains to be had from free speech that make it worth pursuing in this extreme way, by erecting a fortress in its defense.

In the free speech literature there are hints of these perceived benefits. In *Cohen v. California*, Justice John Harlan gave this explanation of free speech:

> The constitutional right of free expression is powerful medicine in a society as diverse and populous as ours. It is designed and intended to remove governmental restraints from the arena of public discussion, putting the decision as to what views shall be voiced largely into the hands of each of us, in the hope that use of such freedom will ultimately produce a more capable citizenry and more perfect polity and in the belief that no other approach would comport with the premise of individual dignity and choice upon which our political system rests.[50]

What is suggestive and interesting about this statement is the idea that the function of free speech is itself to *create* the capacities in the citizens that they are admittedly lacking. How would that educational process occur? Harlan seems to indicate that it results from the choices we exercise over how to use or not use our right to free speech. Surely this is true, but there is another and more compelling way in which it might happen. The educational process may be a variation on Mill's idea that the interaction of falsehood and truth produces a "livelier" belief in the truth. But instead of obtaining a richer appreciation of the true beliefs they already hold, it may be that in the process of having to *defend* their beliefs and ideas, people can be expected to acquire or improve their *capacity* to deal with contrary opinions and beliefs.

This process, however, would seem to depend on the necessity of having to defend their position, and it may be that those afflicted with the tendency to excessive intolerance possess the requisite power that makes that unnecessary. If that is so, or if the educational process breaks down for other reasons, then why might we still wish to pursue the right with such passion? Might the right of speaking give us the opportunity to educate those in need of education? Or is there something else?

In the *Skokie* episode, we have some very revealing testimony to how deep the fortress model can cut through the thinking of free speech proponents in cases at the margins. In chapter 1, it was noted how difficult it is, in thinking about *Skokie*, to feel you have gotten to the bottom of what was going on in that dispute, to feel you know what was really in the minds of the parties and interested observers. For many people connected with the controversy, however, the legal problem in the end appeared to come down to a simple matter of the need to exchange protection of bad speech for greater security for good speech. Beneath the quagmire of platitudes and quotations extracted from the generously large barrel of classical rhetoric about free speech lay this straightforward calculation of expediency.

The most powerful illustration of this is to be found in a book, previously referred to, that was published shortly after the case had been decided by one of the principals in the dispute, Aryeh Neier. At the time of *Skokie* he was the executive director of the American Civil Liberties Union. Neier's book, *Defending My Enemy*, is, as one would expect, intended to provide a fuller statement of the ACLU's position in *Skokie*, but it becomes a very personal statement as well. As an expanded brief of his organization and as an analysis of the doctrinal and theoretical implications of the case for First Amendment protection, the book is unexceptional. As a personal statement, on the other hand, the book is remarkable. Once one wades through the phrases slung right and left about how free speech protection in the case was necessary for the search for truth, for democracy, and so on, one reaches the central, motivating issue for Neier. Symbolically and ironically, that occurs in the prologue to the book, beyond and apart from what constitutes the conventional thinking about free speech. It is there, in the prologue, that Neier speaks not only as the director of the ACLU but also as an individual, as a Jew, and he speaks with a stirring voice. He recites his "credentials" for despising Nazis, recounting his last-minute escape from Hitler Germany as a young boy. "I recite my own background," he explains, "to suggest why I am unwilling to put anything, even love of free speech, ahead of detestation of the Nazis."[51]

From this position he develops his argument, one largely shorn of the usual encrusted rhetoric and platitudes that have so often fossilized our thinking. He appreciates the risk of persuasion inherent in permitting the Nazi ideology to flourish unencumbered by legal restraints: "The risks are clear. If the Nazis are free to speak, they may win converts. It is possible that they will win so many adherents that they will obtain the power to abolish freedom and destroy me."[52] Though he professes that "John Milton's view that truth will prevail in a free and open encounter with falsehood is my view, too," he adds that he "cannot accept Milton's principle as infallible" and is "wary of putting too much faith in any principle of human behavior."[53]

Still, Neier says, he "must examine with care the alternatives that are available to me." The only "alternative to freedom is power." As he tallies the risks and benefits of each alternative scenario, he reaches this sober conclusion:

> If I could be certain that I could wipe out Nazism *and* all comparable threats to my safety by the exercise of power, perhaps I would be tempted to choose that course. But we Jews have little power. We are few in number. We are known by the world as a separate race and a separate religion. Only Jews are doubly marked as a people apart.
>
> The rest of the world is suspicious of us Jews. We are like each other and we will stick by each other, the world believes. If a scapegoat is needed for any evil, look among Jews and accuse all Jews. If a Jew took part in the Crucifixion, all Jews are Christ killers. If a Captain Dreyfus is a traitor, all Jews are traitors. If Karl Marx—despite his childhood baptism—is a Jew, all Jews are revolutionaries. If a Jew lends money, all Jews are usurers. If one Jew is a participant

> in a financial scandal, the Jews are manipulating the economy. Because he is identified as a Jew, the Jew captures attention. There are Jews everywhere. We can be blamed for everything.[54]

Given this reality, Neier is clear in his own mind what legal principles are called for:

> Because we Jews are uniquely vulnerable, I believe we can win only brief respite from persecution in a society in which encounters are settled by power. As a Jew, therefore, concerned with my own survival and the survival of the Jews—the two being inextricably linked—I want restraints placed on power. The restraints that matter most to me are those which ensure that I cannot be squashed by power, unnoticed by the rest of the world. If I am in danger, I want to cry out to my fellow Jews and to all those I may be able to enlist as my allies. I want to appeal to the world's sense of justice. I want restraints which prohibit those in power from interfering with my right to speak, my right to publish, or my right to gather with others who also feel threatened. Those in power must not be allowed to prevent us from assembling and joining our voices together so we can speak louder and make sure that we are heard. To defend myself, I must restrain power with freedom, even if the temporary beneficiaries are the enemies of freedom.[55]

Here, then, we finally have before us the central argument. Protection is not just demanded by the prior cases; nor is it logically compelled by some general commitment to the search for truth or the democratic ideal; nor does tolerance seem the most attractive alternative because of some intrinsic and abstract problem of line drawing in the free speech area. It is, rather, a matter of self-protective political strategy, a response to a perceived reality of ever-threatening intolerance and prejudice by the politically powerful against the politically weak. To such groups, which possess only a fraction of the power needed to secure their social position, a legal principle becomes, therefore, a refuge, but one oddly secured by admitting into it the archenemy. As such, the act of tolerance becomes at once an ambiguous symbol of safety and vulnerability.

Neier's argument states in dramatic form a very common, though partially submerged, line of thinking underlying the operation of free speech. It is not limited to those who are members of groups that have traditionally been so vulnerable to the effects of social prejudice. It is a far more universal feeling. Virtually everyone knows the fear that he or she may become the unpopular member, the victim of discrimination, and from this fear is bred the thought that the wiser course is to secure insurance through a general principle of legal protection. It is a fear not only of the government but of the masses of people that make up that anonymous body known as the public. From this fear of being a persecuted minority the fortress model derives its appeal.

But it is important to bear in mind that the model seeks more than a legal outcome. It involves an entire way of thinking about the notion of free speech. It is, as we have seen, not only a matter of *what* is protected to serve as a buffer but of *how* that protection occurs. The administration of the free speech principle must be made to yield the desired effects of insulation. The primary ingredient in this strategy is the idea that the people should be made to feel the suppression of speech is both unthinkable and unlikely to be successful and that the judges should be led to think they have no choice but to protect the speech if the public should decide to suppress it.

## V

You cannot fault the fortress perspective on free speech, as we could the classical model, either for failing to deal with the extreme cases or for being naive about the characteristics of mass society. It offers a way of conceiving of free speech at the outer perimeter that is comprehensive and unblinkingly realistic. It offers a practical, pragmatic perspective of the world. It is conscious of the threat of conflict within the society and of the need for barriers to keep power from falling into the hands of those who will be inclined to sacrifice freedom for orthodoxy. It focuses, furthermore, on the process of presenting the free speech principle, not just on the particular results to be reached.

There is also, in one sense, undeniable merit and appeal to the fortress-model approach. The idea of limiting choices within free speech doctrine, of making the doctrine as closely knit as possible to avoid the tears and snags that come from the inevitable rough handling it will undergo, makes good sense. If you seek the benefits of speech activity that we have identified in our discussion of the classical model, then there are gains to be reasonably anticipated from erecting the fortress. You have at least preserved something of value, and possibly the activity you have thus preserved will itself become a means of dealing effectively with the larger phenomenon of intolerance, if only by offering the opportunity to notify one's allies of the approaching danger in other areas.

Yet, as we have seen, there are serious risks to the fortress approach. The buffer zone is not always a neutral area but one sometimes filled with the enemy, whose freedom of action is now also preserved. There are fundamental problems that need to be worked out about the relationship between the aims of free speech under the fortress model and the broader human and social ends the classical model has identified for us, as well as about the propriety of the tasks the fortress model places on the shoulders of our judicial institutions.

Probably the most serious cost of the fortress approach is the problem of introducing an unattractive elitist outlook into free speech thinking and analysis. The fortress strategy relies on the abstract language of "the majority," "the public," "the masses," of "they" and "we." As such, it can be a beguiling posture to assume in the defense of the free speech principle. It identifies a more or less amorphous group as possessed of bad tendencies, which will affect us, the innocent few. It can be an overly simple—and, for that reason alone, appealing—account of the world, with evil lurking behind every potential crevice in the doctrinal structure and offering a curious but nevertheless seductive image of vulnerability and superiority. The risk, therefore, is one of alienation between groups. Free speech becomes a divisive force within the community.

It also introduces into our public institutions, however, an unfortunate manipulative frame of mind, a warfare mentality where each side is tacitly setting the rules by which future battles will be fought. One wonders whether the legal institutions of this country ought to be engaged in this kind of implicit bargaining, whatever the merits of the position as a tactical maneuver, instead of attempting to reach for nobler ends. But the courts themselves have participated in this legerdemain. Claims about the history and the language of the First Amendment, about precedents, and about the actual bases of judicial thinking all are often nothing less than misrepresentations. The fortress model contributes to a tendency to rely excessively on a kind of legalistic method of solving social problems, and it is shortsighted about the ultimate ends one can hope to achieve through the free speech enterprise.

These concerns seem sufficiently serious to make us want to consider other ways in which to conceive of the functions free speech ought to serve in the society. While the fortress perspective may not be dismissible as irrelevant to the actual world we inhabit, or as illogical, there may be other and better means of securing the ends we seem to be seeking through the idea of free speech. This desire to consider alternative conceptions of free speech is also spurred by the sense that the fortress model simply does not account for all that we derive from the principle, especially when it is applied at the extremes. Some of us may feel more intensely the risks of being an unpopular minority than others do, though I would assert that all of us more or less have that feeling at some time. But I doubt that the appeasement of that fear accounts for the sense of pride we derive from thinking that in the United States even the most radical and extremist ideas are tolerated; I suspect something more is involved.

Perhaps we can usefully frame the question for ourselves in

this way: If we ever came to feel that the opportunities for our own speech as well as for other speech we genuinely value were totally secure—just as Wigmore argued was true for American society—would we still have good reason for taking the free speech principle to the extreme position we have or would we properly expect the principle to slide into senescence?

# 4
# The Quest for the Tolerant Mind

Each of the two models we have thus far considered possesses only a limited power to explain the modern application of the free speech principle. The classical model rests on premises and assumptions that are either inapposite to many of the problems actually encountered in contemporary free speech disputes or greatly oversimplified in light of the modern understanding of human nature. Focusing as it does on the matter of identifying the primary beneficial uses of speech activity, the classical model has necessarily little to say about protecting speech in the extreme cases, where those values exist only in a most attenuated form. Moreover, its appreciation of the social harm sustained as a result of seriously constricting society's ability to regulate speech activity seems naive at best, and in any event unfairly understated.

The fortress model, on the other hand, offers a strategem that, while making the extreme cases more explicable, can rest on a highly troublesome conception of social reality. In order to erect a legal barrier against legal restraints on speech, it tends to postulate a social universe in which the citizenry is alienated from the government, as well as internally from each other, and to induce a posture that can be unfortunately disingenuous and manipulative. It is also questionable whether it offers an effective strategy for the limited ends sought under it, or whether the social reality it presumes is in fact true.

In their pristine forms, therefore, both perspectives seem seriously incomplete; they fail in substantial measure both to account for what is actually done in the name of free speech and to offer a satisfying conception of what should be done in its name.

At the same time, both approaches contain much that is good and of legitimate appeal. As a statement of aspiration and positive encouragement, as a holding up of an ideal toward which human society should strive—whether it be truth seeking and rationality or personal self-realization—the classical model strikes successfully at deeper cords. In redefining what our ancestors fought for, we remind ourselves of how far we have failed to claim what they have won. On the other hand, the primary strength of the fortress model is that it does not shrink from critical examination of the human character. Rather than extolling our virtues, under its influence we are ready to criticize and to locate defects.

It would seem preferable to blend the best of both approaches, the idealism of the classical model and the realism of the fortress model, while shedding their limitations. Is that possible? If one examines the American experience with its free speech principle in this century, can one identify an emerging conception of free speech that is capable of answering the challenges leveled at the classical and fortress models? Is free speech performing another function besides those thus far considered? I believe it does and has, though for many reasons that function has failed as yet to reach the surface of our free speech discourse.

## I

Let us begin by placing the discussion of this chapter more solidly in context with the development of the argument in preceding chapters. It will be recalled that in chapter 2 we considered, in response to the claims of the classical model, how important the needs were behind the act of intolerance toward speech activity. Seeing these needs as stemming from powerful elements in the human personality, relating to the sense we have of our own identities, we warned against trivializing them. By chapter 3, when we considered the perspective of the fortress model, we found that free speech thinking was, actually, highly cognizant of the needs of intolerance, so much so, in fact, that it tended to view these needs as constituting an impulse that threatened the very foundations of liberty of speech and as therefore requiring a legal check. Instead of minimizing the needs of intolerance, free speech thinking was now seeing them as potentially overpowering, at least for a significant segment of the population. At this point, however, we began to wonder whether this vision had erred too far in the other direction by overstating the danger of excessive intolerance for the liberty of speech and, in any event, to wonder whether a legal barrier was the most effective means of dealing with the problem, given that the impulse to intolerance would seemingly manifest itself in many ways other than through excessive legal restrictions on speech activity. These difficult questions raise complex empirical issues about social reality and strategy, but now I would like to suggest that it is possible to accept, at least partially, the line of argument or premises that have been developed up to this point and yet find a crucial social role for the free speech principle in the context of the assumed reality of an impulse to intolerance.

To do that, we must begin by readjusting our vision of the social role of free speech, partly by altering our vision of what functions law can and does play in the society (in fact, by borrowing from the conception of the functions of law used earlier in developing the case for permitting regulation of speech activity). We might accept (to a substantial degree, if not entirely) the fortress model's characterization of a tendency in the society to react with excessive intolerance toward speech *and* the answering claim that that tendency will inevitably manifest itself well beyond the free speech context—and yet find the free speech principle a worthwhile enterprise precisely because it is one very important means by which the society attempts to deal with that larger problem. In this conception free speech derives its appeal by providing a method of addressing a ubiquitous social incapacity, not just a means of securing protection from that bad tendency for the special activity of speech, perhaps with the added hope that the insulated speech activity will somehow itself solve the larger problem.

At this stage, however, it would be better if we described the purpose behind the principle not as that of protecting speech but rather as that of dealing with the phenomenon of what we have called the "impulse to excessive intolerance" generally, though we do that by insisting on an extraordinary degree of toleration only in the limited context of speech activity. The role of free speech is directed at developing a capacity of far greater moment than that of just regulating the appropriate level of legal restraints on speech activity in the society. The legal principle operates in a small sphere, and in a special way, in order to address a larger issue. Law (in this case, *constitutional* law) is being used not simply as a barrier against entry but as a major project concerned with nothing less than helping to shape the intellectual character of the society.

In the remainder of this chapter we shall examine how such a vision of the social principle of free speech would work and, in particular, how it might assist us in understanding the three fundamental issues that have so bedeviled the development of a viable free speech theory for modern times: Why should we exercise such extraordinary self-restraint in the regulation of speech when we do not with respect to nonspeech behavior? Why, in particular, should we tolerate extremist speech? Why should we vest the interpretative and enforcement functions of the principle in the judicial branch? More immediately, however, we must return momentarily to a point developed in chapter 3—that excessive intolerance of speech is only a particular manifestation of a more widespread bias in social behavior—in order to lodge more firmly in place what is actually a cornerstone premise of this new perspective. With this argument, as with so many others we have now considered, we will actually be constructing a new theory of free speech out of many of the same materials we have so far used in uncovering the limits of the classical and fortress models.

We understand that we can be "intolerant" in many ways and in many areas of social interaction. But we do not always see the common roots of much of what we classify as intolerant behavior. In part this is because of the way in which we approach thinking about the free speech rationale. We look at speech for the good it can bring, not for the harm it may cause or the responses it may generate. As we observed in chapter 2, traditional efforts to justify the special protection afforded speech under the free speech principle have tried to identify those prized goals thought to be especially realizable through the activity of speech—goals like the exchange of information and ideas and personal self-fulfillment. Speech is defended as integral to such pursuits, and the regulation of speech as a threat to their realization. While, of course, accurate to a degree, this general perspective has the unfortunate consequence of limiting our understanding of the possible social roles for the idea of free speech.

At the extremes it makes free speech proponents appear to give excessive weight to these human pursuits, to the neglect of other important human ends (as Wigmore observed). More important, by focusing on those activities that can be pursued through speech in order to define the social justification for free speech, we tend to pay little attention to the thinking that lies behind the general human response to speech acts. We see it only as an enemy to be fought off, while we seek other ends like the acquisition of information. By neglecting it, we tend in turn to isolate its social presence, both in the sense of seeing how broadly it cuts through social interaction (both for desirable and undesirable ends) and in the sense of seeing how it involves feelings that are problematic for everyone (though the context in which they arise may be different for different people). We ought to begin, therefore, by exploring some basic propositions about the dimensions of the bias sometimes exhibited in the act of censorship.

If it is a tendency of human nature to overreact in the use of legal restraints against speech activity, we must expect that tendency to manifest itself in the form of nonlegal coercion as well. That there is a wide network of social controls for speech activity is indisputable. All the time we face the problem of deciding whether and how to respond to people who say things we believe are wrong or who communicate messages in contexts we think are inappropriate and hurtful. Should we refuse to associate with the speaker, shunning him or her in the way that some religious communities do with enormous effect? Should we ridicule the speaker, casting contempt on him or her—thereby using, and in the process proving, the potency of speech to inflict injury (just as we discussed in chapter 2)? The possibilities are only as limited as our ways of expressing disapproval. The power to affect our opportunities for employment, social life, and status in the community is all within the hands of the established majority.

In fact, free speech law itself recognizes this reality, though in a somewhat backward fashion, for both the common law torts of libel and invasion of privacy, which free speech jurisprudence has thus far tolerated (albeit with some modifications), permit civil actions by individuals against those who have accused them of holding or advocating certain beliefs.[1] Free speech cases have also held that a state may not eliminate the opportunity for anonymity in public debate and association, recognizing that a requirement of public disclosure of one's identity will inhibit many people from expressing their views because of the power of social stigma.[2] In these ways, First Amendment law acknowledges the personally devastating consequences that can sometimes befall individuals when the community comes to believe or know they have held and advocated certain beliefs. To have it said that you were once a communist sympathizer, a fascist, an atheist, or a liar can make you, at least in most quarters within the society, socially and economically a pariah, as destitute as if you had been thrown in prison and fined. In fact, a good case could be made for the proposition that the power of social intolerance exceeds that of legal intolerance, as Mill argued:

> [I]t is the opinions men entertain, and the feelings they cherish, respecting those who disown the beliefs they deem important which makes this country not a place of mental freedom. For a long time past, the chief mischief of the legal penalities is that they strengthen the social stigma. It is that stigma which is really effective, and so effective is it that the profession of opinions which are under the ban of society is much less common in England than is, in many other countries, the avowal of those which incur risk of judicial punishment.[3]

Regardless of the relative power of nonlegal and legal sanctions, we may reasonably assume the same tendency to employ them excessively, which we acknowledge with respect to legal coercion, must exist for nonlegal coercion as well. There would seem to be little reason to think that people's attitudes toward legal restraints and punishments are peculiarly prone to be unrestrained.

What of our reactions to behavior other than speech? Is a similar tendency to overreact or to react intolerantly experienced there as well? The answer, of course, is yes. All the time we face, both in our individual and in our social encounters with people, the problem of deciding just what beliefs or values we will demand that others conform to. As to these choices, we ought to expect that if the manifestation of thoughts through speech acts produces in us feelings that can generate improper reactions, then the manifestation of mind through other, nonspeech conduct will do the same. And if we look at speech acts from the traditional libertarian perspective—that speech is simply a prelude to action—we should expect to find that the responses evoked by speech will undoubtedly also be evoked by the action itself. In fact, this logic is widely accepted, if only tacitly, by traditional libertarian defenses on behalf of the free speech idea. Mill's classic argument for tolerance, for example, was general:

It envisioned a basic impulse to establish orthodoxy with respect to all kinds of behavior and then attempted to provide a universal principle by which it could be said that a certain category of behavior (Mill's idea of self-regarding behavior) would be entitled to near total toleration. Speech was simply one subspecies within that category. That type of inquiry into the possibility of finally deriving a general principle by which we can define the appropriate boundary for social control of behavior continues to this day, still focused largely—as it has been for so long—on the matter of sexual conduct.[4]

There are indications even within the jurisprudence of the First Amendment itself that point up the need to guard against the impulse to intolerance in areas beyond speech activity. In religion, for example, the protections afforded by the First Amendment cover not just the advocacy of religious viewpoints and beliefs but the practice of religion as well. The impulse we are considering does not discriminate between speech and other acts; on the contrary, it concerns itself with any manifestation of the underlying attitudes and beliefs and, so, must be guarded against in all contexts. Moreover, an entire area of nonspeech conduct, referred to as "symbolic speech" (basically nonspeech behavior that communicates ideas—a part of the First Amendment we will take up in more detail in chapter 6), has been grafted onto the First Amendment to protect it against the same impulses that seek to control speech activity. Dress, for example, is a form of nonspeech behavior that, as Bagehot remarked, may stimulate the desire to persecute.[5]

The problem we face, therefore, is not simply one of controlling an impulse to insist too strongly on our beliefs and values with respect to speech; the impulse threatens all behavior. Everyone who is perceived as being different, as having different values or beliefs or an interest in a different way of life, is a potential victim of an excess of this impulse. Besides members of religious groups, the most common victims of such intolerance are those of different races or nationalities. We commonly refer to this form of intolerance as "prejudice" rather than "censorship," but it is usually stimulated by the same underlying psychology. What leads us to react with intolerance is, typically, a concern with the *mind* perceived to be at work—with the way of thinking of the person or persons, whether that be political beliefs or general attitudes or values or whatever one might call it; and, equally important, with the fact that this thinking is essentially being communicated by the actions of those who hold, or appear to hold, these different beliefs, attitudes, or values.

Although it requires that we reconceive somewhat our notion of the term *ideology,* it may be useful to think of intolerance in all of these areas—toward speech and nonspeech behavior; political, religious and racial groups; aliens and foreigners—as motivated in large part by a concern with the ideology, or way of thinking, manifested by the victims' behavior (which, unfortunately, for some victims means simply being). We tend not to see this reality with respect to something like racial prejudice because the line commonly used as the basis for triggering the intolerance (skin color) is less precise than that used for other intolerance (for example, against "communists") and because the ideology *perceived* to exist, and thought of as troublesome, is so much broader and less specific than the usual political or religious ideology that commonly prompts intolerance. (Again, communism may serve as an example of a more concrete political ideology and Catholicism of a system of religious belief.) Perhaps another source of our failure to make these connections is the structure of the Constitution itself, which treats the problems of excessive intolerance in these various areas as discrete, primarily by enunicating different principles and by putting them in numerically separate categories. Nonetheless, the fact remains that the intolerant responses dealt with in these various areas of social intercourse bear an important unity of underlying thought or motivation. It is instructive, in this regard, to recall that Neier, when he was trying to defend the ACLU positions in *Skokie* on fortress-model grounds, cast the specter of potential intolerance against Jews as involving not just invasion of speech rights but all forms of anti-Semitic prejudice. The history of deep intolerance against speech is usually linked to excessive intolerance against groups of people, especially aliens (the most infamous of censorial statutes is the Alien and Sedition Act).[6]

Up to this point, in considering if the feelings that lead us to want to censor (improperly) speech acts also lead us to want to prohibit or sanction (again improperly) nonspeech acts, we have really been focusing on a particular kind of problem—namely, deciding whether behavior we dislike or disapprove of should be sanctioned or prohibited *at all*. As we have seen, if we are moved to intolerance by the verbal manifestations of certain thinking, we are almost certainly going to be moved similarly by nonverbal actions that reflect or implement the same thinking. As we have noted, many people, following in Mill's footsteps, have struggled with this problem of trying to identify the line at which social control of behavior is appropriate. While some areas of human activity (such as religion) are now widely regarded as beyond the legitimate control (at least legal control) of the community, there is a range of activities (sexual practices, for example) about which there is wide disagreement concerning the proper degree of social regulation. The issue of demarcating the line between the legitimate social regulation of morals and the sphere of private liberty is, therefore, still very much with us and undoubtedly will continue to be so for the foreseeable future.

But it would be a grave mistake to think that the problem of dealing with what we have been calling the impulse to excessive intolerance is restricted to deciding what we should choose to prohibit. Of greater importance—certainly of greater importance to the meaning of free speech we are developing here—is the problem of deciding *how much* to punish behavior that is legitimately within the realm of community control. The problem we face with the impulse to excessive intolerance, in other words, is not just with resolving the issue of what merits sanction but also with answering the question of how to punish that which is properly punishable.

From the time that Paris seduced Helen and ignited the Trojan

War (and Agamemnon insulted Achilles by taking his concubine), people have written and wondered about how concededly bad acts can threaten and challenge the identity of individuals and communities and pose the deepest issues for them about the proper course of action to take in response. Today we can see the problem vividly and repeatedly presented to societies as they struggle with the problem of how to deal with political crimes, particularly acts of terrorists. A striking example for this country was the attack on the American embassy in Iran a few years ago, which resulted in the extended holding of American hostages. Like all guerrilla and terrorist attacks, this one had an explicit and conscious communicative purpose, a violent act of propaganda. A wave of anger swept across this country, posing a serious danger to visiting Iranians, who were sometimes violently attacked in retaliation for the actions of their distant countrymen. Though the anger in this country at the Iranian government was clearly justified, and equally clearly justified *some* intolerant response, it also ran the risk of becoming seriously excessive and of leading to the injury of innocent individuals.[7]

Of paramount significance for our purpose is that this anger stemmed from precisely the same concerns that Holmes identified as constituting the foundation of intolerance (or, as he put it, of "persecution") toward speech activity and that we discussed in chapter 2 when we considered the range of potential harm from speech acts. The Iranian act posed a challenge to the *identity* of the United States. It raised a question of what in fact the values of this country were and, more pointedly, a question of whether the United States had the courage and the power to act on those values. It was an insult, a threat, a challenge; just like any of the verbal variety, it expressed a way of thinking to others and "set an example" for others to follow. The situation, in short, involved the same dynamics as that posed for the Jews and for American society generally by the Nazi march in Skokie, Illinois, or, indeed—to carry through another thread from chapters 2 and 3 that we shall develop more fully in chapter 7—as that posed for free speech proponents by the attempt to forbid the

Nazi march. That one was a "speech act" (though, it should be noted, the "speech" was manifested in a variety of forms of conduct: placards, dress, walking and gait) and the other an "act of violence" involving direct physical harm may be relevant for some purposes, but it does not alter the reality that the fundamental issues of gauging the right response, and of addressing and controlling the same important inner feelings that arise because of the act, were present in both instances. The need to do something in the face of the attitude—or the ideology—revealed, and the problem of deciding just what that something should appropriately be, were present in both instances. The same was true, to provide additional examples, of the Russian invasion of Afghanistan, which, of course, prompted the subsequent Olympic Games boycott, and of the asserted Cuban activity on the tiny island of Grenada, which prompted an armed invasion (though, interestingly, these may also be seen as sequels to the Iranian crimes).

At this point it is perhaps useful to pause and recall that, in chapter 2, we in fact have already made this observation about the similar difficulties presented for observers and the community by speech and nonspeech acts. There, however, the point was part of a plea that we become more sensitive to the harm-producing capacity of speech acts. Now, building on that recognition, we are able to see in the similarity of the dilemmas posed by such acts the comparable difficulty of resolving those dilemmas successfully. Out of this observation we will in due course see how it provides the basis for a justification of free speech as we have come to know it.

The dramatic international events we have so far mentioned only highlight an important common element in confrontations with behavior deemed improper—namely, that every bad act contains an implied threat to repeat the act, as well as other, more serious acts, and is, in a sense, an example for everyone to follow. In this way, then, all behavior (verbal and nonverbal) is "communicative," or, what is really more to the point, all behavior reveals the thinking of the actor, which in turn poses the same issues for the community. This is because, ultimately, we are always concerned with the mind behind the act itself, whether it involves speech or nonspeech behavior. "An evil act need not be repeated and can be repented of," Tolstoy wrote, "but evil thoughts engender evil acts."[8] While such a concern with the mind of others is reasonable to a degree, the problem is that it poses a constant and serious risk of getting out of control and of generating undesirable responses.

So the point really reduces to a simple one: Just as the problem of setting and enforcing the appropriate limits of speech activity within the society is rendered difficult by feelings basic to human nature (involving belief and identity), so are our responses to nonspeech behavior rendered difficult by the same feelings. That there may be more social justification, or greater social need, behind the regulation of nonspeech conduct (perhaps, ironically, because of the greater effectiveness of nonspeech forms of communication—a point we will return to in chapter 6) than of speech conduct is not the relevant issue at the moment. What is true and important for the present discussion is that the system for dealing with nonspeech conduct must take into account, and develop checks and protections against, the very same tendency to excessive intolerance that manifests itself in the speech context. Aristotle made the point long ago: "It is not everybody," he said, "who can find the centre of a circle—that calls for a geometrician." So, too, it may be said, it is "easy to fly into a passion." Anyone "can do that." But "to be angry with the right person and to the right extent and at the right time and with the right object and in the right way—that is not easy, and it is not everyone who can do it."[9]

And so, for example, we should find, as we do, in the system of criminal justice various protections against the impulse (in Holmes's words) to "sweep away all opposition." The decorum of the courtroom, the presumption of innocence, the rules of evidence restricting the submission of evidence likely to be emotionally inflammatory (discussed in chapter 2), and the cautions surrounding the meting out of punishment are all intended to function, at least in part, to check the impulse of the various decision makers to give vent to their desire to punish excessively someone who has potentially put into action a way of thinking deemed wrong by the society.[10] For, inherent in the act is a mind at work that the community must necessarily feel to some degree is a challenge to its beliefs, its identity. The mind behind the act puts those who come to know of it in the same psychological dilemma as does the racial epithet or the advocacy of wrongful behavior. The acts of conviction and punishment are the society's communicative response, which is now labeled "deterrence" instead of "censorship."

For all its vast social significance, the system of civil and criminal norms is actually only one of the many arenas of social interaction where the feelings behind the impulse to excessive intolerance emerge as an important element to be dealt with. Since the impulse involves nothing less than an internal conflict, within the individual and the community, over how to respond in settings in which people advocate and act on different attitudes, values, or beliefs, we readily see how pervasive to social interaction are the feelings we are talking about.

The feelings must arise and must be controlled in the basic operation of a self-governing political society, where a willingness to compromise and a willingness even to accept total defeat are essential components of the democratic personality. Democracy, like literature, it may be said, requires a kind of suspension of disbelief. At the norm-setting level, as well as at the enforcement level, a capacity to contain one's beliefs in the interest of maintaining a continuing community is critical. The problem of deciding on the nature of the commitment to one's belief is one of exquisite complexity. Those who possess the power to see their choices put into effect must decide whether and how far to press ahead in the face of opposition.[11] Those in the minority must decide whether to accede to the will and power of the majority or in what ways to continue the fight, perhaps by selecting a course of action from the various modes of resistance or, in extreme cases, by resorting to outright rebellion. No bright markers serve to guide these subtle choices. In this sense, therefore, the capacity sought through free speech bears a special relevance to the actual functioning of a democratic system of government.

The same capacity cuts even more widely through the social structure and can be observed at various levels of social interaction. Consider the basic requirements of a functioning bureaucracy and of various social professions.

A central injunction to those who make up the bureaucratic organizations of government is that they must implement the orders and directives of others despite their own contrary personal beliefs and feelings. Obviously, we draw limits on this; a bureaucrat or an employee should not act contrary to law, and a considerable gray area exists where personal integrity is both respected and encouraged, or even morally demanded. But, on the whole, a basic ethic of the employee's position is to set aside his or her personal beliefs and to implement those of others. The same is true, perhaps to an even greater extent, with certain professions, most notably that of law itself. Again, keeping in mind the complexity of the interaction between the lawyer's sense of right and wrong and his or her obligation to defend the client, there remains at the core of the lawyer's role an obligation to represent the positions of others even though the lawyer's own beliefs and values may differ from those being defended. Law students are taught to "think like a laywer," which primarily means learning to see that on every issue there is "another side" to the one the student is so inclined initially to believe is right. Legal institutions in this country could not function unless these skills were developed; but developing them is no easy task and the impulse to violate the ethic remains powerful. (It is not, incidentally, without interest that the principal argument advanced in favor of the lawyer's ethic of submerging personal beliefs and values is that advanced in the free speech sector—that one's faith should be placed in the system, and not in oneself, to arrive at the proper accommodation of competing interests.)

The foregoing observations, though perhaps elemental, are nonetheless pivotal in arriving at an understanding of the relationship between tolerance of speech activity under the free speech principle and of behavior in other areas of social interaction. As noted earlier, if we believe that the sort of excessive intolerance we encounter and fear toward speech acts stems from conflicts we have over the beliefs and values we hold and profess allegiance to, it would seem that something so basic to human nature should certainly be expected to arise in a host of other areas in our lives, since beliefs and values are put at risk in so many social settings. If all this is so, if the issues evoked by confrontations with speech acts are characteristic of a host of social encounters, we might find it helpful to think about the principle of free speech and about the functions it serves in the light of that common feature—perhaps as one means by which self-control of those feelings is sought. Rather than isolating the activity of speech to understand the social role of "free speech," we might find it more helpful to see free speech as intended to help us gain control of feelings relevant to the whole tapestry of social intercourse. To that inquiry we now turn.

Perhaps it would be best to begin with a statement of the general theory of free speech that emerges from the preceding discussion. The theory proceeds from the premise that the way of thinking behind intolerance of speech activity (in the form of legal coercion) is something we encounter throughout social intercourse. It is not peculiar to legal coercion against speech activity, or even to coercion against speech. It is universal, both in the sense that it is something everyone must learn to control, though in varying degrees, and in the sense that it is pervasive in any society organized along the lines of twentieth-century United States. We might say that when we deal with speech acts, we are dealing with the mind behind the act, or revealed through the act, but to say that is to begin to recognize that we are virtually always concerned with the mind as manifested through behavior, whether it be speech or nonspeech behavior. The universality of the issues confronted in the speech context means that tempering this way of thinking can be a principal focus, a primary source of value and justification, of the free speech principle. Stated another way, it need not be the exclusive function of free speech to protect an activity (speech) that is especially prized, but rather to work toward the correction of a perceived general defect in our intellectual makeup, one that happens to manifest itself sometimes in our reactions toward speech acts.

A central function of free speech, therefore, is to provide a social context in which we collectively speak, in a public and official setting, to an important aspect of what we might think of as the intellectual character of the society. Taking this approach, we can now see that the purposes of the free speech enterprise may reasonably include not only the "protection" of a category of especially worthy human activity but also the choice to exercise extraordinary self-restraint toward behavior acknowledged to be bad but that can evoke feelings that lead us to behave in ways we must learn to temper and control. What is important about speech is not that it is special but that the excessive intolerance we sometimes experience toward it is both problematic and typical, in the sense of reflecting a general tendency of mind that can potentially affect many forms of social intercourse.

At the beginning of this chapter, as well as before, it was said that any sound theory of free speech must answer three central questions. The first is to explain why the society has chosen to abide by what amounts to a presumption against regulation of this one area of behavior, that is, the behavior of speech. Apart from a few isolated regions in the area of nonspeech behavior, nowhere else in life do we insist in this way on such a level of self-restraint. We now have a thesis on that issue. From the critical insight into the generality, the universality, of the feelings that generate an excessive response to speech acts, we can see free speech as a limited, or partial, area in which an extraordinary position of self-restraint is adopted by the society as one means of developing a more general capacity with respect to that impulse. Free speech provides a discrete and limited context in which a general problem manifests itself and in which that problem can usefully be singled out for attention. Does it make sense to structure social life in this way?

Envisioning free speech from this perspective leads us to think about how society gains by structuring itself in nonlinear ways. Given the hold that the notion of "equality" has on our minds, and especially on the legal mind, this may not at first seem a comfortable prospect from which to examine the functions of a legal doctrine. Yet it is critical that we do so. For when we examine social life, rather than finding perfect consistency between all its various parts—a uniformity of treatment smoothed out through the careful evaluation of relevant differences—we often encounter a functional world composed of discrete inconsistencies, or anomalies, which together, however, can form a coherent whole. Such a method of behaving, therefore, is not so anomalous as it may at first appear.

In both our private and public lives we often act in certain ways in one area in order to influence our behavior in other areas—in ways that, considered alone and independently from other parts of our lives, would appear strange. We understand, if sometimes only instinctively, that qualities we acquire by stretching ourselves in one area of activity spill over into other areas where those qualities are also needed. We may even separate certain areas of life where we will behave in extraordinary ways in order to develop particular qualities needed elsewhere. The near total restraint in the use of race as a consideration in public decision making, which functions in part as a symbolic counterbalance to the use of race in private decision making, is a good example of this.

This bending over backward, pushing ourselves to an extreme, often in a limited context, is common in life. We parcel out our world in this way perhaps more than we recognize, certainly more than we acknowledge in the area of legal principles. We do so because it offers advantages over a homogenized existence: It can provide a more manageable place in which to begin, where success is more likely and, if achieved, will in turn strengthen future resolve. It also has the great merit of allowing us to limit the costs of our mistakes, which can have a liberating effect, permitting us to experiment more freely and to take risks that would otherwise be inhibiting. It also allows us to focus attention with greater clarity on the problem being addressed, whereas if it were treated generally, it might get lost amid a sea of other problems.

The attractiveness of a limited and an extraordinary effort is even greater when the capacity we are trying to achieve is beyond the reach of the usual methods of social control—when what is ultimately involved is a matter of attitude and capacity for searching introspection.[12] When that is true, the selection of a single area, in which there will occur a public and perhaps rather rigid and indiscriminate rejection of the type of thinking deemed improper, can offer an important means of symbolizing the "proper" way of thinking, which, it is hoped, will then be employed throughout all areas of behavior.

Furthermore, to select a discrete area in which to develop a particular quality desired can have the advantages of concrete experience, offering the kind of practical knowledge that comes from actual experimentation. If, for example, we are concerned about being unduly risk-averse, we may benefit by exposing ourselves to significant risks in a limited context where the consequences of those risks are relatively contained. In this way we may begin to learn how we overestimate risks and how to make adjustments to modulate that tendency in other settings.

Finally, in chapter 2 we saw how actually taking the step of passing a *law* prohibiting certain undesirable behavior is usually a much more powerful demonstration of a community's commitment to the values threatened by that behavior than is a simple verbal declaration of such a commitment. So, too, may an official *act* of extraordinary self-restraint toward that behavior similarly provide a more forceful demonstration. The injury voluntarily sustained helps to establish, both to oneself and to others, the depth of one's feelings and the purity of one's motives. The punishment willingly accepted by the person who engages in civil disobedience, or the injury passively suffered by those choosing a course of nonviolence in response to injustice, and the real and significant consequences thereby accepted help to prove, as well as inspire, commitment and to establish that good motives underlie the behavior.

All this speaks importantly to the functions served by free speech. Speech activity under the operation of the principle becomes a discrete area of behavior that is treated in such a way as to give it broader social meaning. The general problem addressed is one that cannot be solved by adopting a position of total abstinence, as we are sometimes able to do with troublesome impulses. The feelings that generate excessive intolerance cannot be obliterated from social interaction, nor would we want them to be. Like the desire for personal gain, the impulse of intolerance must be controlled and channeled toward good social ends, not uprooted. Moreover, the problems we face are not susceptible to ordering by some general rule or test. We are dealing with a matter of attitude, of balance and control. Under these circumstances, then, because of the feeling that the process of decision making is infected with a potential bias in the direction of intolerance, we can see the plausibility of stretching the capacity of tolerance in a single area where the bias can be expected to make an appearance. To structure our social life in this way helps achieve the desired end. The deprivation suffered has its necessary and unavoidable costs, but it also means—because of the distorting power of the bias over our judgment—that there will often be significantly smaller costs than our estimates would lead us to predict, thus providing a useful practical lesson in the importance of employing a corrective factor when we are called on to react to similar feelings in other contexts. The self-restraint in the face of the injury sustained demonstrates powerfully, more powerfully than a general injunction to be appropriately tolerant in all circumstances ever would, to ourselves and others, a commitment to exercise moderation throughout social intercourse. In this respect, the operation of free speech may be seen as reflecting a determination to create a general intellectual character through the creation of a kind of tolerance ethic.

We may now, finally, also see why the category of speech behavior seems instinctively such a sensible place in which to undertake this enterprise. That speech generally causes less individual and social injury than does nonspeech behavior, while not in itself a sufficient justification for a free speech principle such as we now have, is nonetheless an important characteristic for explaining why speech is an appropriate setting in which to pursue a greater capacity for tolerance. Speech offers a fairly sharp line for limiting the extraordinary experiment with tolerance and an upper limit of potential for harm that makes unrestrained activity there generally tolerable for these purposes. There is in this sense, then, rational support for the widespread feeling that the reduced harm-producing capacity of speech behavior is somehow relevant to understanding why the culture has developed different rules for the regulation of speech and nonspeech acts.

But what of the extremes of speech? Why pursue the principle to this extent, to the very outer perimeter of speech activities, to speech nearly all of us believe immoral and vicious? This is a more complicated question, requiring a more lengthy response, though it is a response that emerges clearly from the vision of free speech we are now considering.

Let us take a quick survey of the possible functions of extreme cases and then consider them more fully in the specific context of *Skokie*. First, if one has decided to construct a partial response to a general problem *as a means of dealing with the general problem* (and not just with the problem in the limited sphere selected), an insistence on an extreme degree of self-restraint is helpful as a means of emphasizing that generalized meaning. Extreme cases may be useful in realizing the desired symbolic significance. The extremes tend to attract attention, and that, as any educator or radical knows, can be pedagogically and symbolically advantageous. They provide, in other words, a useful context in which to impress on people the lesson sought to be communicated, a frame for the message. But there is more to the extreme cases than this.

Given that the very nature of the problem we are dealing with is, as a practical matter, beyond the capacity of law to solve, involving as it does what is essentially a matter of attitude or judgment, it seems reasonable to approach the problem in the limited area selected for symbolic action through a principle of nearly general self-restraint. It is simply too difficult to make a case-by-case examination of legal restraints on speech to ascertain whether the underlying motivations are of an improper variety. The problem of the impulse to excessive intolerance is simply too elusive for that type of scrutiny.

Even apart from the complexity of the inquiry in the free speech area itself, there is the additional point that the problem of the impulse—because it cuts through a variety of social interactions and involves a capacity for toleration in the broadest sense—must really be confronted by creating something of an *ethic* against regulation, which will exert force in the opposite direction, very much like the presumption of innocence does in the context of the criminal jury trial. One way to accomplish this is to hold the society to a position of near complete and total tolerance in a limited area of social intercourse. Therefore, not only does the context of the extreme case provide a desirable educational setting for conveying the general message; it also means that the society has committed itself to a course of action in which the sacrifice demanded will create a psychological environment in which the message will find its most receptive audience. By pursuing the "principle" to its logical end, well beyond what the particulars of individual cases call for under it, the society impresses on itself the importance of the lesson, creating out of it an ethic, or identity, of self-restraint. "To straighten a bent stick you bend it back the other way."[13]

Through this process, moreover, the society may learn more

about its capacity for self-restraint. The difficulty of untangling the mental processes behind restrictions on speech behavior (or, for that matter, wherever the problem arises) also means that actual self-restraint may offer the benefits of real-world experience. If we think we are too inclined to be risk-averse, too prone to overstate the risks involved in social interaction, or too cautious about permitting differences to exist, a benefit of going to extremes is that it provides a good basis for testing our grasp of reality.

This observation leads to another. Thus far, one may have thought the discussion has assumed that the extreme cases are largely free of the problem of improper intolerance but provide a useful context in which to create an effective counterbalance to the problem as it might arise in other contexts. This assumption is untrue, or at least needs to be seriously qualified. If one looks at the speech involved in extreme cases, the usual conclusion is that the speech itself is without serious social value and, considered independently, may properly be forbidden or restricted. But if one looks not at the speech but at the motivation behind the restrictions, one may properly conclude that the restrictions were imposed for bad reasons. In the typical extreme case, the wrong motivation tends to be treating the speakers as scapegoats.

Finally, it is also the case that we often benefit from the extreme cases by holding up to ourselves, through the act of protecting the speech and permitting it to occur, the very example of the mental process that it is the fundamental purpose of free speech to alter. Extremist speech is very often the product or the reflection of the intolerant mind at its worst and, as such, an illustration to us of what lies within ourselves and of what we are committed, through the institution of free speech, to overcome: Perhaps ironically, but nonetheless powerfully, the principle of "free speech" serves to "protect," and so to hold up before us, that which we aspire to avoid.

Again, *Skokie* is helpful in the role of general illustrator. Clearly, the speech activity involved in that case pressed the principle of free speech to an extreme degree. Attention was focused on the case as a result, and a dialogue on the controversy over the proper limits of toleration spread throughout the country, not all of it, of course, of high quality but a surprising amount was (including a television drama about the case).[14] The case provided in effect a forum in which attention was drawn not only to the limits of free speech but more broadly to the problems and issues of the prejudice of anti-Semitism, in all its various forms. What gave the case its dramatic power was not that it involved making a pact with the devil to better secure one's own freedom, but rather that it involved a confrontation with the more complex, and less comfortable, processes at work behind the desire to punish these speakers, whether by legal or nonlegal means.

The wish to prove to oneself and to others that one has "no doubt either about one's power or one's premises" is, as Holmes said in his peculiar way, "perfectly logical." It can be extremely unsettling to see our own bad qualities reflected in the behavior of others; it draws our attention to what we regularly close off, or censor, from our minds. But such unsettling feelings can turn into abhorrence when we see those bad qualities under intense magnification, when we thus have put before us the potential course of those bad tendencies we sense within ourselves. Then the desire to dissociate ourselves from the behavior is proportionally intensified. As much as we may wish it were otherwise, and as painful as the acknowledgment is, each of us bears some aspect of the character of mind reflected by the Nazis. And it is the intolerance we feel toward our own intolerance that contributes to our wanting to censor the external, exaggerated reflection of that part of ourselves. In its extreme form the desire manifests itself in a violent response, such as we regularly witness—usually with shame and embarrassment—in the confrontations and threats of confrontations of those who challenge the speakers. Some, like the Jewish Defense League (JDL), attempt to turn this type of behavior into a virtue by making it appear as if only they have the courage to stand up to evil.[15] But the desire may also take less primitive and more official forms as well.

The extremist groups are, of course, well aware of this potential for uncontrolled responses and shape their behavior so as to be as provocative as possible. Sometimes this is accomplished with considerable subtlety, which, of course, only adds to the provocation. When we began examining the *Skokie* dispute, and at subsequent times, I observed that the case involved a hall-of-mirrors effect. We always felt uncertain about precisely what was going on. Here again we can see how that feeling can arise. One element of the confusion was over the messages the Nazis would be conveying. The original petition described the purpose of the demonstration as a protest against the restriction that had been placed on their ability to hold a rally. It therefore became a demonstration against the denial of free speech rights—on the surface, then, many steps removed from a demonstration to advocate genocide, racial hatred, or the institution of a fascist dictatorship. Yet, the location selected for the march, the clothing and insignia they intended to wear, the movements of their bodies during the march, as well as the ideological position they ultimately wished to advance once they secured their free speech position, all bespoke another set of messages. In essence the desired effect was to "clothe" their position in innocence—in this case, ironically, in the drapery of the First Amendment—and in doing so to make the behavior provoked appear all the more indefensible.

It should be noted that one of the courts in the *Skokie* litigation actually gave some hint of recognition of a correspondence between speaker and audience in the *Skokie* context. At the very end of the federal court of appeals' opinion, when the judges returned to making a personal statement about the difficulty of drafting such an opinion, they expressed their "extreme regret" that "after several thousand years of attempting to strengthen the often thin coating of civilization with which humankind has attempted to hide brutal animal-like instincts, there would still be those who would resort to hatred and vilification of fellow human beings. . . ."[16] What is particularly interesting about this statement is its unusual willingness to recognize something of the commonality between the Nazis and the rest of us. The internal intincts referred to are not simply those of an unrefined and uncivilized group but part of the condition of everyone, for which civilization offers only a "thin coating." However much one may disagree with the assessment of the human condition in the words chosen, the judges should be praised for not having censored an unpleasant truth at stake in the case. It was unfortunate, however, that the truth was offered only by way of denunciation of the plaintiffs and not taken as the potential basis on which to build a rationale for the decision itself.

If one tries to trace the sources of the intolerance impulse still further, we find it becoming even more potentially complex. Beneath the surface rhetoric in the dispute about the risks of persuasion and offense, there rests the possibility of a quite tragic undercurrent of emotional energy at work in the confrontation. The Jews are victims, whether they were survivors of concentration camps or not. The horrifying treatment they received at the hands of the German regime must be regarded by all as one of unspeakable inhumanity. But the injuries of that event are far from simple. They consist not only of the loss of parents and relatives and friends or the pains of persecution and the fear of its revival. They also include a way of thinking about the event, of which guilt no doubt potentially plays a significant part. We cannot here explore this proposition at length, and in any case, the observation has been so frequently made and discussed by others that for me to do so would be redundant.[17] But in its essential outlines it involves a feeling of not having resisted sufficiently, of having survived when others did not, and of having in some obscure ways, perhaps through weakness or some affirmative behavior, contributed to the event one abhors. That such feelings have no basis in reality is beside the point, for they are real feelings and as such can serve as springs of behavior.[18]

For purposes of thinking about free speech, that behavior may be the desire for an excessively intolerant response to a group like Collin's Nazis. Such a group both provokes the issue of guilt and provides the opportunity for its denial—in effect, for its censorship. Their very vulnerability, moreover, adds to the invitation.

In *Skokie* a distinct line of thought was sometimes heard, though it did not surface explicitly in the judicial opinions, that bears on all this. It was usually expressed as the argument that if we permit ourselves to prohibit this march by the Nazis in Skokie, then we will be forced into the position of having to permit whites in the South to forbid civil rights marches by blacks, for they too find such speech "offensive."[19] This was, of course, an argument for the free speech position in *Skokie*, and it was an argument of considerable power. But its real power lay not in pointing up the problem of effectively drawing lines (for, as I argued before, it is unconvincing to claim that we could not distinguish between a Nazi protest and a civil rights protest), or even in the acknowledgment of the difficulty of rationally explaining differences between cases, but in a strange and elusive identification between the claims for suppression that might be made in both cases. What made the cases appear comparable, I think, was the similarity of an improper or undesirable motive potentially involved in the controversy. Just as Jews might like to censor internally the guilt of the victim, and we all might like to censor the guilt of the piece of Nazi within us, so, too, many whites in the South have their own guilts to appease. It would, therefore, not be a simple matter to face Southern whites and explain that one's own intolerance was grounded in proper considerations while theirs was not. The point is not that intolerance in Skokie could not have been implemented for the right reasons, only that it might not have been, and it is that potential for improper intolerance that makes an explanation very difficult.

One must be careful here not to be misunderstood. As said before, the point is not that the guilts many Jews may feel as victims are justified in reality, nor that Jews necessarily feel this way or were motivated by it in the *Skokie* case. And it is certainly not being said that the Jews in Skokie were in any way other than in a certain psychological sense potentially similar to the imaginary white bigots who played such a powerful image in the thinking about *Skokie*. Rather, the idea is to see that there are bad as well as good reasons for being intolerant toward groups like the Nazis in Skokie. The point, therefore, is that in the extreme cases there is the *potential* for a type of internal censorship to be present, which makes them important contexts for confronting the problems the free speech principle is concerned with addressing.

There are, then, at least three primary advantages that can arise in the context of social confrontation with extremist speech. An initial consideration that is important, though perhaps not significant enough in itself to warrant tolerance of extremist speech, is that it is useful for gaining the notice needed for the concept to play its larger, symbolic role. Like a monument designed to inspire the virtue of courage, an extreme and greatly enlarged version of toleration can better enlist our attention.

Two deeper considerations are at work in the typical extremist speech case, one leading into the other. The first is that extremists often represent the paradigm of the intolerant mind at work and thus illustrate the very qualities of mind that it is the purpose of the free speech principle to counteract. In this, I do not mean to refer, as many people do, to the fact that many extremists hold or advocate anti–free speech views. Rather, I am talking about the more general qualities of mind revealed: their attitudes toward their own beliefs and those of others, their incapacity to cope with uncertainty in human affairs, and their quest for simple and clear answers—of which their attitude on liberty of speech is but a piece. For these people, speech is a weapon of intolerance, and through their behavior they represent what through the principle of free speech the society is committed to avoid.

Just as conferring awards for good actions is a means by which a community creates its own identity and defines its values, so, too, it is possible to accomplish the same ends by the opposite means of holding up before itself that behavior the community most seeks to avoid. The energy to achieve a desired state of mind or behavior can come from either direction; and with free speech we follow the direction of Montaigne:

> There may be some people of my temperament, who learn better by contrast than by example, and by flight than by pursuit. This was the sort of teaching that Cato the Elder had in view when he said that the wise have more to learn from the fools than the fools from the wise; and also that ancient lyre player who, Pausanias tells us, was accustomed to force his pupils to go hear a bad musician who lived across the way, where they might learn to hate his discord and false measures.
>
> The horror I feel for cruelty throws me back more deeply into clemency than any model of clemency could attract me to it. A good horseman does not correct my seat as does an attorney or a Venetian on horseback; and a bad way of speaking reforms mine better than a good one. Every day the stupid bearing of another warns and admonishes me. What stings, touches and arouses us better than what pleases. These times are fit for improving us only backward, by disagreement more than by agreement, by difference more than by similarity. Being little taught by good examples, I make use of the bad ones, whose lessons are common. I have tried to make myself as agreeable as I saw others violent.[20]

Often in extremist speech cases there is a similar phenomenon at work. Though it is ironic in conventional First Amendment terms, it is nevertheless true that extremist speech may sometimes produce a chilling effect on less explicit varieties of the attitudes communicated. The extremist speech makes us perhaps more conscious of the full potential that lurks behind seemingly more innocent versions of the same ideas.[21] With some ideas, this may be highly desirable.

It would be a serious mistake, however, to think that all that occurs in the protection of extremist speech is a healthy recoiling at the magnification of the intolerant mind. Confrontations with extremist speech are usually beset with more complications than that. The mistake would be to think that these sorts of cases do not really involve the potential for the type of excessive intolerance or overreaction that it is the function of the free speech principle to confront. The truth is the contrary, for it is very often the case in these types of controversies that the most excessive responses are entertained and acted on. The very "puniness" of these people makes them a convenient and an attractive group on which to demonstrate those wishes that Holmes talked about as constituting the roots of "persecution," and their extremeness provides a rationalizing pretext for venting underlying angers.[22] The magnification phenomenon only accentuates the felt need to dissociate oneself through the act of intolerance, and the speakers thus provide the ready makings of a scapegoat.

Finally, to gain emphasis through repetition, it should be said again that I am not suggesting that intolerance of extremist speech is always infected with improper considerations or motives. Rather, the contention is that intolerance in these cases *may* be done for bad as well as good motives, even when the restrictions or punishments imposed are not, viewed independently, excessive. As with everything, good things may be done for the wrong reasons, and with free speech the reasons are what matter most. We ought to be concerned, for example, if Nazis were banned as a kind of denial that anti-Semitism is a latent force within the society.

In the end, all this comes down to a simple but nonetheless critical point: extremes are not to be understood as the peripheral cost of an inevitably imperfect world, in which no one can be trusted to draw the proper lines properly, but rather as integral to the central functions of the principle of free speech.

The third and last major theoretical issue to be considered is whether it makes sense in a democracy to entrust the interpretation of constitutional norms to the judicial branch. This is a matter frequently discussed in the legal literature. Typically, proponents of judicial review point to the desirability of having some bulwark against an overreaching majority—a bulwark favored with the power of a government institution, yet independent of the usual pressures of popular government.[23] I do not wish here to rehearse those arguments or to try to resolve that general debate, but I would like to identify some generally unnoticed advantages of a judicial enforcement system for a principle of free speech such as we have so far outlined in this chapter. It should be borne in mind that the tolerance principle, as I have thus far described it, is intended and designed to perform a self-reformation function for the general community and not, as is so often assumed as the starting point for discussions about the functions of judicial review, to offer a shield of protection either for the majority against the government or for minorities against unfair treatment at the hands of the majority. In the light of that deeper social function, what can be said about the benefits of having the free speech idea enforced by a judicial branch?

The free speech principle depends for its success on being able to provide an organizing principle for, or an account of, the act of toleration. Without such a coherent and articulated explanation, the confusion and uncertainty about motives that make intolerance the preferable alternative to tolerance will continue to create pressures for legal restraints on speech activity. Keeping this need in mind, we can identify several advantages of a judicial system of enforcement for the free speech principle.

First, the limited number of voices who pronounce the reasons for tolerance (in written statements) makes a coherent explanation more feasible. Furthermore, it may be regarded as important that judges are members of a professional group whose central intellectual ethic is identical with the aspirations of the free speech principle. To a degree more than with any other group, judges are expected to have mastered the tolerant mind, to have the capacity to set aside their personal beliefs and predilictions, and to control the impulses that accompany them as they set about interpreting and administering the society's laws—as the judges in *Skokie* observed repeatedly, it may be recalled.[24] A good part of Holmes's great reputation as a judge, for example, was due to his capacity to separate his own personal beliefs from his judical obligations. Though, here as before, important complications surround this central judicial ethos, both as to the full desirability of the model and as to its actual achievement in practice, it is important that we do not let these complications obscure its fundamental role in defining the nature of the institution. The ideal judge, in short, is in important respects a paradigmatic illustration of the qualities of mind pursued through the free speech principle, and it is therefore appropriate that we should assign to judges the task of safeguarding and administering that principle.

There are two final attributes of the judicial system that, while difficult to identify or even articulate, are nevertheless of potentially great significance to the actual functioning of free speech. The first involves the fact that the judiciary, having no forces at its disposal for enforcing its judgments, is the one government institution most dependent on its own capacity to secure the "toleration" of others necessary for its acts to become effective.[25] This dependence on others may well induce a greater institutional understanding and appreciation of the process of toleration in general.

The second attribute arises from an ambiguity, an uncertainty, under the free speech principle over the responsibility for the choice for tolerance. We observed in the analysis of the impulse to intolerance how it arises, at least in part, from the feeling that the choice of tolerance would indicate weakness of belief or will. Such a feeling may be especially strong in the extreme cases, which we now can see are integral to the new free speech principle. The sense, therefore, of having no choice in the decision for tolerance, and thus no part of the responsibility, can help to alleviate the need to act intolerantly. The ambiguity about this ultimate decision maker is very much at the center of the constitutional and judicial system of enforcement of the free speech principle. The "judge," a figure not subject to election or recall, appears to be the final decision maker, and the text interpreted (the Constitution) is itself encased in a set of social attitudes that makes its revision virtually unthinkable. The practical consequence of such a system is that in particular cases, when the needs of intolerance run extremely high, there is in the structure of free speech a means for making tolerance more palatable, which thus helps to preserve the identity of the general principle. Simultaneously, the litigation setting offers a convenient context for those for whom tolerance is troublesome both to articulate their "plea" for intolerance and to dissociate themselves from the result, and thereby from at least part of the "responsibility." (They must, of course, still decide whether to obey the judicial order, and to that extent tolerate the speech, but what is typically regarded as an important new ingredient—respect for the system of law generally—has now been introduced into the overall calculus.)

On this last point, it may be noted that we have already observed, especially in the *Skokie* cases, an identical process reflected in the behavior of judges themselves. When the strain of tolerance gets too great, when the beliefs and values underlying the law conflict with the deeply held beliefs and values of the judge, we often find the judge turning to a posture of choicelessness, typically by disclaiming any power to alter the "predetermined" legal result—a phenomenon Professor Robert Cover has so powerfully documented for us with respect to the behavior of Northern pre–Civil War judges in their enforcement of the fugitive slave laws.[26]

It is significant, therefore, that the First Amendment has an ambiguous status in terms of social consent. In a sense, it is "ours" and reflects what we claim to want; but in another sense, it is "beyond" our control and even "imposed" on us. This ambiguity about consent and the First Amendment creates a beneficial environment for the discussion and implementation of it. The *process* is well designed for the problem intended to be addressed. The process of judicial enforcement gives those opposed to the speech an institutionalized method of objecting to the speech, and thus at least partially the opportunity of signifying their position with respect to it, and a result (if the judicial decision is in favor of toleration) that is—partially or ambiguously—"beyond their control." Judges, therefore, serve the function of taking over the decision to permit the speech to go forward or not, and in so doing relieve those who object of a share of the responsibility. (This is similar, perhaps, to the way in which official and judicial enforcement of the criminal law provides a beneficial psychological distance for the public, which is able to give expression to feelings of outrage and indignation while a more deliberative process handles the prosecution and punishment of the offender.) The upshot is that free speech, in the way it operates in fact, can serve to make tolerance more possible than if it were left to be just a general injunction in the usual legislative process.

I would close this section with two comments. First, I am not claiming here that the posture of choicelessness is either entirely good or entirely satisfying to the intolerance impulse. I am saying that in a system that calls for extreme tests of the capacity of toleration, the system is aided by being somewhat ambiguous on the score of who exactly is making the choice for tolerance. On the other hand, the context must also permit some embrace of the result, for otherwise no gain in a general identity could be had, at least not one with any integrity to it. "Our" Constitution, with its method of judicial enforcement, permits both results.

The essence of successful institutions is often to be found in their having room for subtle and ambiguous solutions to difficult problems, and this would seem to be so under the institution of free speech. We need to look further than we have into the structural meaning of the First Amendment. In chapter 7 we will return to the problem of defining the judicial role in free speech cases.

## II

It is accurate to say, as the fortress model does, that there is present in human society a grave and difficult problem of obtaining toleration between the minds within it. But it is also accurate and vitally important to see two other fundamental aspects to this impulse: First, the human feelings involved are socially problematic not only with respect to legal restraints against holding or advocating certain beliefs—matters with which the principle of free speech is immediately concerned. Second, the problem of controlling those feelings is universal; that is to say, we *all* must address this problem more or less in our various social roles. It is not an affliction of only a few, or even of a majority. Provided we add to these acknowledged features of the impulse the assumption that we are still educable on the basic problem, we have the basis for a new theoretical perspective on free speech.

In chapter 3 the idea was introduced that the impulse behind intolerance toward belief and speech is really but a particular manifestation of a larger phenomenon and social problem. It was raised there as a basis for criticism of the fortress perspective, a basis for the suggestions that the general incapacity thus revealed in the discussion suddenly cast into doubt the processes of democracy and truth seeking, which had long been the mainstays of free speech theory; that it accentuated the risks of extremist speech, which had to be counterbalanced against the gains we could expect to derive from protection of that speech; and that it threatened to overwhelm and render useless the primary gain itself, namely, the legally secured right of speaking. But it was also doubted whether the risk assumed by the fortress model is really as significant or substantial as the model supposed it to be, and we wondered, in any event, whether there would be a role for free speech if the risk were in fact not so great.

Now, in the fortress model's insight into human behavior we have seen another basis for defending free speech: Free speech may be an institutionalized method of grappling with the larger problem that involves our attitudes toward our own beliefs and identity. Through the social process organized under the principle, the society highlights the larger social problem; by confronting the problem in that discrete context, the society acts not just to protect one area of behavior against the force of the impulse but rather to address the whole problem. When libertarian theory bemoans the reality of a tremendous, historically verified tendency toward excessive intolerance, it is reasonable to ask why we haven't established some institutions that would be concerned with dealing with the bad tendency, in a sophisticated, educative way, and not simply by checking its impact in a minor area. In American society, free speech may have come to be one such institution.

Part of the reason for this development may be due to a phenomenon that the fortress model implicitly recognizes and uses to its advantage, namely, that free speech is more than simply a result; it is also a social *event*. People assemble in courts, they present arguments for and against the relevance of the principle, officials listen to the arguments and then arrive at a decision they must defend and justify according to standards developed over many years. The process we have created for implementing free speech involves a means of discussing and defining the bases on which we protect speech, and therefore the social meaning we derive from it. It is, in a very real and practical sense, a *forum* within the social life in which we have the opportunity to shape and affect our intellectual attitudes and, through that, our behavior.

For free speech to assume this function, however, it is critical that the feelings we seek to control, that constitute what we have referred to as the impulse to intolerance, be seen not only as cutting through a broad range of human behavior but also as being universal to all members of the society. For without that acknowledgment there is both a disinclination to address the problem openly, as well as sympathetically, and an inclination to slip into the posture of superiority toward those perceived to be afflicted by the defect. Human nature being what it is, shared experience usually provides the basis for empathic and therefore meaningful discussion. When Neier set out to justify to Jews and others why the Nazi march should have been protected in Skokie, he prefaced his argument with the statement that he too was a Jew and a near victim of the concentration camps. Holmes often spoke confidently about the necessities of a country at war while noting that he himself had been a veteran of war and, better yet, a casualty. Such examples could be repeated endlessly. The acknowledgment of shared experience, or commonality, makes it easier to address troublesome issues and concerns. It lessens the sense in the listeners' minds that the speakers do not understand or have minimized their feelings and that the speakers regard themselves as superior, using the occasion to criticize as an opportunity to deny that they personally share the defect criticized. It therefore facilitates, instead of defeats, communication.

With the recognition of the wide dimensions of the impulse to intolerance, it becomes possible to see the enlarged social relevance of the free speech principle. To arrive at this point, we must reconstruct the meaning of the principle by amalgamating and expanding some of the basic premises of the classical and fortress models, in part through exploring the defect that the latter model identifies and through retaining the sense of hope for improvement that the former offers. Together, the two models reflect an appropriate degree of ambivalence on a vital social issue.

## III

And so we see the elements of a new and different perspective on the meaning of free speech, one we may refer to as the general tolerance theory. In this view the social function of free speech is to provide a focus on the mind behind the act of intolerance rather than to protect the activity of speech itself as something that possesses independent value.

To see free speech as concerned not just with protecting the activity of speech but with the reaction to that activity, and to the personal values reflected in those reactions, changes considerably our idea of the ends served by the principle. We have already seen the range of importance to the community of learning to exercise self-restraint toward behavior found offensive or threatening. It seeks to induce a *way of thinking* that is relevant to a variety of social interactions, from the political to the professional. Significantly, this perspective sees the social benefits of free speech as involving not simply the acquisition of the truth but the development of intellectual attitudes, which are important to the operation of a variety of social institutions—the spirit of compromise basic to our politics and the capacity to distance ourselves from our beliefs, which is so important to various disciplines and professional roles.

It also promises a benefit we all can feel, individually as well as collectively, of avoiding the burdens that the impulse to intolerance can impose on us, or that through it we impose on ourselves. To escape its demands or, more accurately, to reduce the power of its grip, to become the master of the fears and doubts that drive us to slay the specter of bad thoughts, is an achievement of the first magnitude. This is what those I quoted earlier, like Meiklejohn and Holmes, who associated fear with intolerance and fearlessness with tolerance, may be read as saying. The idea was captured more explicitly by Brandeis in his opinion in *Whitney v. California*, among a litany of phrases about the purposes of the First Amendment: "Men feared witches and burned women," he wrote, and it "is the function of [free] speech to free men from the bondage of irrational fears."[27] Brandeis recognized the importance of self-control and that it brings benefits not just to those who have been spared as victims but also to those who have been spared the costs of acting badly toward them. In the case of those who burned witches, the gain would have been far more than the trivial saving of their not having to gather wood; it would have involved their being freed from the agony of fear and guilt that must have initiated and accompanied their acts of brutality.

It is now possible to answer the charges by Wigmore, and others who have wondered as he did, that free speech in the extreme cases is just the product of an excessive infatuation with liberty and freedom, a loss of balanced judgment on the needs of society for order and decency, or a perverse misuse of the social commitment to truth seeking and democratic government.

For under the general tolerance function, free speech is not concerned exclusively with the preservation of a freedom to do whatever we wish, or with the advancement of truth or of democracy as those terms are generally used, but with the development of a capacity of mind, with a way of thinking; it is concerned with facing up to a perceived bias of mind, one that interferes with all of those objectives, as well as others, and is also encountered in the decisions over the regulation of speech. Wigmore was right about the tendency to overstate the risks to freedom and to understate the needs of society for intolerance, but he—like many others—failed to see the importance of free speech in helping control the impulse that may lead us to "smite on the mouth" those who go among us advocating and acting on views we dislike, even legitimately so.[28]

Free speech in this sense, then, is not inconsistent with what we do toward speech in other areas, like juries. There we seek means of controlling the impulse just as we do with the principle of free speech, though in each case in a different way—one by limiting speech and the other by tolerating it. Concern over a troublesome (but natural) way of thinking is what unifies our behavior in each area.

In many respects the principle of free speech is a typically American response to a perceived problem. Whereas theorists like Mill have attempted to fashion a single marker by which to gauge the justice of all social reaction to every form of behavior (speech or other), this society has evolved an approach with no single, overarching rule to follow but with a mixture of granting and withholding power, a kind of check and balance, which seeks in an untidy way to arrive at a workable compromise on a matter deemed—correctly, I think—beyond the reach of "rules" to order for us.

It is this that most fully accounts for the strength of the free speech principle's appeal for this society. It would seem to be what we have in our minds when we say in defense of free speech in a difficult case, it is "the principle that matters." The "principle" is the choice to exercise extraordinary self-restraint toward injurious behavior as a means of symbolically demonstrating a capacity for self-control toward feelings that necessarily must play a role throughout social interaction, but which also have a tendency to get out of hand. In this secular context we derive something of the same personal meaning and satisfaction of the religious fast, a self-initiated and extraordinary exposure to temptation that reaffirms the possibility of self-control over generally troublesome impulses. This concern with feelings that cut broadly through social life leads us to think about free speech as our premier constitutional principle, of wider compass and significance than all the others.

In chapter 5 we shall explore how this notion of the meaning of free speech can be found in the classical literature on the subject, though you must look carefully for it and even then you will often see the evidence only as shards sticking through the surface of the discussion. The reasons why it should have remained so submerged are not difficult to identify. First, the language of the First Amendment speaks of prohibiting the "abridgement" of "freedom of speech," and it is only natural that our attention should therefore focus on figuring out why that activity possesses sufficient value to deserve that sheltered status. Second, it is common in legal thinking generally to treat what seem to be extreme cases as of marginal importance, understandable only by reference to their relation to a core of most valued examples of the activity. Third, the legal mind is primed by training and practice to see explanations in differences, not similarities. Fourth, we have inherited a rhetoric about free speech that speaks in terms of earlier functions, in terms of the "liberty" of each individual and of the community of citizens against the sovereign state.

Perhaps the major intellectual obstacle to appreciating the shift in social meanings of the free speech idea is an implicit reluctance to see that ideas can change their social meaning over time and, further, that an activity taken to an extreme can also change its meaning from that it possessed when pursued in moderation. One might usefully think of the meaning of free speech as an evolutionary process with three basic stages. In the first stage the people are attempting to secure basic political power and rights for themselves, as at the time of the American Revolution. In such a period people genuinely need protection against governmental misbehavior. In the next stage this battle has largely been won with respect to the government, but not as against the people. Society itself retains a perceptible tendency to sacrifice the essential elements of a democratic system of government.

In the final stage the basic elements of the governmental process are secure—including, most importantly, the element of valued speech activity. The society has achieved a certain stability, in which its core values are reasonably clear and widely accepted. Yet there remains a broad range of social functions in which a way of thinking is both necessary and desirable and yet in some doubt. Here the focus shifts importantly away from the form of behavior (speech) and toward the mental process behind the reaction to that behavior. The greater the security of the behavior, the more internal and introspective the self-examination can become. But the mental process that gives cause for concern is present both in the activity of speech and in the activity of intolerance toward that speech. A choice is available for proscribing it in one or both, but the choice is made to select one.

Here it becomes a matter of importance that the mental process is present not only in the interaction with speech but elsewhere, too. The toleration of speech then assumes its larger, symbolic function. The meaning of extremes is, in part, to serve that function. At precisely that point, the meaning of what is done in the name of free speech changes.

# 5

# The Internal Dialectic of Tolerance

By examining closely the thinking of two major First Amendment figures—Alexander Meiklejohn and Oliver Wendell Holmes—we can illustrate and extend the general perspective on free speech arrived at in chapter 4. In the First Amendment writings of these two men, we see powerfully represented how, beneath the epiphenomenon of traditional justifications for free speech, modern theorists have been led to focus less on the worthiness of speech activity as a basis for protection and more on something potentially problematic in the public response to speech acts.

Instead of seeing the purpose of free speech as simply that of facilitating decision making by the removal of impediments to the flow of information and ideas, or that of maintaining the area of speech behavior as a special preserve for individual self-realization and fulfillment, both Meiklejohn and Holmes found that free speech provided the occasion for thinking about the general character of mind as it was reflected in the public efforts to suppress speech. The potential of the way of thinking that motivated the censorship troubled them and guided their responses, just as the potential of the way of thinking reflected in the speech act contributed to the desire to punish the speaker. But the remedies each offered were themselves in deep conflict, and we will see a profound ambivalence about the basis and meaning of tolerance.

We will try here, then, to uncover what each of these free speech theorists diagnosed as the central intellectual problem of their time and what each proposed as an intellectual antidote to that problem, in the special context of free speech. In doing so, we will find that the high clarity of each man's vision, together with the sharpness of the contrast between those visions, makes them especially forceful examples of an internal conflict within free speech thought and within the posture of tolerance generally. Identifying and articulating an acceptable answer to the feelings underlying the impulse to intolerance turns out to be no simple task. Here we encounter two important modern attempts to construct a self-image for tolerance, one an image of self-doubt and the other an image of affirmation of belief. Each represents something universally felt, though each is fundamentally at odds with the other.

In the preface to his book *The Negro and the First Amendment*,[1] Harry Kalven observed that the idea of free speech was marked by an unusually keen "quest for [a] coherent general theory."[2] Every area of law, Kalven puzzled, was rife with inconsistency and ambiguity; yet inexplicably there was little tolerance for anomalies in the field of free speech.[3] As to why this was so, Kalven speculated that "free speech is so close to the heart of democratic organization that if we do not have an appropriate theory for our law here, we feel we really do not understand the society in which we live."[4]

Kalven's unusual sensitivity led him to see both the singular quest for a unified theory of free speech and the sense of a close relationship between understanding free speech and understanding "the society in which we live." In this chapter we shall consider both the nature of that broader relationship and at least one significant reason why no unified theory has been forthcoming.

We begin with Meiklejohn.

## I

It will be recalled from chapter 2 that with *New York Times Co. v. Sullivan*[5] Kalven thought that a "coherent general theory" had been initiated.[6] In Kalven's view, Justice William Brennan's opinion for the Court had for the first time provided our free speech jurisprudence with a "central meaning," identifying a core function for free speech rather than simply repeating, as the cases had so often done, the "clear and present danger" test of Holmes, a phrase devoid of theoretical explanation.[7] Kalven thought the Court was essentially pursuing the theory Meiklejohn had put forward in the late 1940s.[8] Though the *Sullivan* opinion nowhere mentioned Meiklejohn or his work, Justice Brennan later acknowledged, circumspectly to be sure, the connection.[9]

This triumvirate of *New York Times Co. v. Sullivan*, Harry Kalven, and Alexander Meiklejohn, which was formed in the mid-1960s, spawned a major idiom for talking about freedom of speech and press. Meiklejohn became a cardinal theorist, and many who came after him professed to be following in his path. Professor Robert Bork, for example, began his work on the First Amendment with a statement of general allegiance to Meiklejohn's interpretation,[10] and Professor Alexander Bickel also claimed to be working within this tradition.[11] Whether these self-proclaimed disciples were, in fact, ultimately true to the faith is an open question, but the identification is sufficiently strong to allow the observation that *Sullivan* initiated a major theoretical and idiomatic avalanche, which Meiklejohn was thought to have precipitated.

Meiklejohn's seminal essay is a curious and deceptive work, one not readily categorized or summarized. Its power is elusive. Though the line of argument seems simple to follow, the reader often thinks the real concerns remain implicit. Surprisingly, much of the "logical" structure of the argument crumbles on the slightest handling, leaving us puzzled about how to account for the essay's reputation and impact. To reach the source of its strength, we must descend through several layers of argument. The descent is worth making, however, for it illustrates the rich interplay of functions of the modern First Amendment.

The conventional reading of Meiklejohn cites him for advancing the basic proposition that in a society that has chosen to be self-governing, the government cannot label some information as too dangerous or wrong for that society to hear. To permit such censorship is to negate the original premise of self-government, because to be self-governing, a citzenry must have the information and ideas necessary for decision making. Therefore, Meiklejohn is noted for arguing that the meaning of free speech must be that all speech relating to the process of self-governance is "absolutely" protected against governmental suppression.

Two other corollaries of Meiklejohn's thesis are frequently noted. The first is his claim that the First Amendment protects *only* speech relevant to the process of self-government; other speech is relegated to the reduced protections of the due process clause of the Fifth Amendment.[12] His working concepts for these different categories are what he called "public" and "private" speech. Public speech is that used by people to plan together for the general welfare. Private speech is more selfishly motivated and is not directed at solving public issues: "There are, then, in the theory of the Constitution, two radically different kinds of utterances. The constitutional status of a merchant advertising his wares, of a paid lobbyist fighting for the advantage of his client, is utterly different from that of a citizen who is planning for the general welfare."[13]

The second corollary commonly referred to concerns the scope of permissible regulations within the category of public speech. While such speech cannot be prohibited on the ground that it is "dangerous," it is not, on the other hand, entirely free from any restriction whatever. Meiklejohn claimed to take the practical view. His central image is the New England town meeting, at which people "assembled, not primarily to talk, but primarily by means of talking to get business done." Rules of procedure are therefore necessary and under them a speaker could be declared "out of order," to which it would be no valid response to assert a right to "talk as he pleases, when he pleases, about what he pleases, to whom he pleases."[14] The limitations on speech that Meiklejohn would impose in the context of public decision making arise from the general principle that "the point of ultimate interest is not the words of the speakers, but the minds of the hearers." Or, expressed in another form that has frequently been quoted: "What is essential is not that everyone shall speak, but that everything worth saying shall be said."[15] The principle of free speech in this sense serves a public, or collective, and not a private, or individual interest.

Now, it is difficult to understand why Meiklejohn's essay should eventually hold such power and meaning for people, if all one looks at is this cold, abstract statement of his thesis. Certainly the basic idea, that a democracy requires freedom of expression among the citizens, is, if not obvious, then hardly original with Meiklejohn. One can find in the literature before Meiklejohn many similar statements about this basic purpose of free speech.[16] As to the corollaries, the first—that free speech protects only public speech, as Meiklejohn defined it—has never really caught much attention as a doctrinal principle. The other—that free speech must be interpreted according to what best serves the collective or public interest—again was hardly new with Meiklejohn. That he was prepared to make this the only purpose advanced by the First Amendment may be of interest, but only marginally so. The source of the essay's appeal, in other words, seemingly cannot be in the originality of its central propositions.

Nor can it lie in its logic or method of derivation of those propositions. Indeed, Meiklejohn's methodology is quite unconvincing. This is especially evident in a series of arguments from the constitutional text Meiklejohn uses to support his theses. He argues, for example, that no public speech, however dangerous, can be excluded from public debate. He supports this position in part by making an exegetical foray into the language of the Constitution. First, he notes that Article I, section 6—the speech and debate clause[17]—provides "uncompromising" and "absolute" protection to the speech of our political representatives. Second, he posits that under the theory of democracy, all rights and powers of the representatives are merely "derivative" of the citizens.[18]

The argument is deficient on at least two grounds. First, it assumes away the problem by proceeding on a debatable interpretation of Article I, section 6. But more to the point and more problematic is the unexamined assumption of a perfect equivalence between the speech protected under that clause and the speech protected under the First Amendment. It would hardly seem self-evident that the framers believed the scope of protected debate on the floor of the Congress would be precisely coextensive with that throughout the country at large; indeed, one interpretive stance would be to draw precisely the opposite conclusion from the different language employed in each section of the document. Even apart from what the framers intended, a good case could be made for not regarding both provisions as coextensive. The nature of the process of selection for those who are likely to inhabit the congressional halls might lead one to tolerate much greater freedom of discussion there than among the general population, where indirect controls over irresponsible speech are far less extensive. Only by ignoring the possibility of different degrees of concern about legislative and public debate could Meiklejohn so facilely reach the conclusion that the First Amendment protects, for example, subversive speech because it is coextensive with the speech and debate clause.

Nor would a view opposed to Meiklejohn necessarily negate the choice to be "self-governing," because people continue to retain the right to elect their representatives and to vote on issues of their choice. The problem is one of determining the degree to which, under the Constitution, the general society is to govern itself or, to put the matter slightly differently, of determining just how the process of self-government should be structured.

In sum, the difficulty is in defining and justifying what we mean by self-government and why self-government excludes restrictions on political speech, a problem that Meiklejohn assumes away by this textual legerdemain.[19]

One could go on in this vein, but it would only be repetitious of the central point. On the level of standard "legal" analysis, of putting together a logical interpretation of a text, Meiklejohn's work borders on the unsophisticated. In a review of Meiklejohn's book, Chafee took him to task for attempting to enter a debate for which he was ill-equipped. Chafee recalled an incident at Brown University (where Meiklejohn had been in the philosophy department) in which Meiklejohn had attempted to break up a fight between students and had received a dented hat for his efforts. Chafee cited the event as an appropriate analogy with Meiklejohn's attempt to enter the ongoing debate about the meaning and limits of free speech. It is true that Chafee had little reason to be generous to Meiklejohn, since he was one of the principals attacked in Meiklejohn's essay, and his suggestion that Meiklejohn should not have entered the debate because he was untrained smacked of professional arrogance; but for conventional legal analysis Chafee gave Meiklejohn the correct marks.

Doubts about the magnitude of Meiklejohn's achievement, however, go far beyond questions about his originality and skill in conventional legal analysis. They go to the fundamental issue of whether what he had to say has any real meaning or message for the world we actually inhabit. Here there are two fundamental problems with the analysis, as described so far, both of which we discussed in earlier chapters. The first is whether the theory has anything to say about speech restrictions that are the *product* of the democratic system, and not simply the imposition of censorship by a government acting outside of democratic procedures. The second is whether the theory has anything to say about why in particular we should protect speech that seeks to undermine the system itself—for example, by advocating the violation of concededly legitimate laws. To read the essay with these questions in mind leads us to a deeper structure in Meiklejohn's argument.

Meiklejohn was clearly aware of the reality of democratic intolerance and clearly intended the protection to extend to subversive speech. He was writing in the immediate post–World War II years, at the beginning of the Cold War. The period bore a close parallel to the climate of severe intolerance during and immediately after World War I, when Holmes first dealt with the subject of free speech. At the very beginning of the essay, in the preface, Meiklejohn spoke to this reality of intolerance, noting that an extensive system of internal security had been devised, with widespread public support, to uncover "un-American" and "disloyal" activities and agents. Referring to Federal Bureau of Investigation activities, he said:

> And that procedure reveals an attitude toward freedom of speech which is widely held in the United States. Many of us are now convinced that, under the Constitution, the government is justified in bringing pressure to bear against the holding or expressing of beliefs which are labeled dangerous. Congress, we think, may rightly abridge the freedom of such beliefs.[20]

These were "wretched days of postwar and, it may be, of prewar, hysterical brutality."[21] The question to be answered, then, was should the society refrain from employing legal coercion against these subversive ideas, against those who would say that "the Constitution is a bad document," "that [war] is not justified," "that [conscription] is immoral and unnecessary," "that the [political systems] of England or Russia or Germany are superior to ours."[22] Even more pointedly, he asks, "Shall we listen to ideas which . . . might destroy confidence in our form of government," from those who "hate and despise freedom to those who, if they had the power, would destroy our institutions?"[23] Should we permit the publication of Hitler's *Mein Kampf*, of Lenin's *The State and the Revolution*, or of Engels and Marx's *Communist Manifesto*?[24]

Thus, given this assumed reality, Meiklejohn would ultimately have to confront not only the general question "What do you do when the public has restricted speech through the democratic process?" but also the further question "Why protect this particular brand of speech?" At the time Meiklejohn wrote, not only was there a form of self-government actually at work, though perhaps not the one Meiklejohn would have wanted, but the type of speech being excluded from the system was that which allegedly directly challenged, or sought to undermine, the system itself. His paradigm of the traditional town meeting gave him no obvious solution to these difficulties, for it was not the "chair" alone that was seeking to silence opinion; and it was not obvious why a community could not decide that speech that sought to undermine the system established for the resolution of disputes should not be tolerated.

Why should a theory of free speech, which envisions its purpose as serving the system of self-government, lead to the conclusion that a self-governing society cannot choose to prohibit speech that advocates the end of self-government itself? This was the central challenge put by Wigmore in 1920[25] and more recently by Bork:

> Speech advocating forcible overthrow of the government contemplates a group less than a majority seizing control of the monopoly power of the state when it cannot gain its ends through speech and political activity. Speech advocating violent overthrow is thus not "political speech" as that term must be defined by a Madisonian system of government. It is not political speech because it violates constitutional truths about processes and because it is not aimed at a new definition of political truth by a legislative majority. Violent overthrow of government breaks the premises of our system concerning the ways in which truth is defined, and yet those premises are the only reasons for protecting political speech. It follows that there is no constitutional reason to protect speech advocating forcible overthrow.[26]

Quite clearly, Meiklejohn was of the view that the only interest cognizable under the principle of free speech was the public, or collective, interest and not any individual interest of the speaker.

What that collective interest consists of, especially in cases involving extremist speech (such as the advocacy of a dictatorship), is far from clear if approached in conventional First Amendment terms.

It is either inaccurate or too simple to suggest that Meiklejohn, like Mill, believed the advantage would be a new truth or a livelier appreciation of a received truth. Despite his frequent claim that his was a practical, businesslike orientation to the First Amendment, Meiklejohn never really said what important information would be the payoff for the act of tolerance. Plausible arguments might, of course, be imagined. It might be thought useful to know what everyone is thinking within the society, especially those segments that are most likely to act destructively. As we noted in chapter 2, such data could conceivably be helpful in organizing the society, distributing benefits, or identifying and correcting problems. Or, as we also saw earlier, a more limited argument might be that the society has, through the act of suppression, effectively rendered itself incapable of engaging in self-government because it no longer has the minimally necessary information to make intelligent decisions.

But Meiklejohn never advanced such arguments. Despite his early, rather vague comments that the function of free speech is to help the society get its business done, by ensuring the acquisition of information, the real theme of the essay turns out to be much larger. As the discussion progresses, we find that free speech provides the occasion for making a general assessment of the intellectual character of the society, as revealed by the act of speech suppression, and the opportunity to make a recommendation for an equally general remedy. It is here that the essay's power rests: not in its logic or its abstract concepts, but in its characterization of the meaning of the act of tolerance and its counterpart of intolerance. The "theory" of free speech is not really theory in the usual sense of the word (an abstract description of the world as it is or ought to be), but part of an overall rhetorical effort to persuade us to become the sort of people Meiklejohn would like us to be.

Meiklejohn's concern, therefore, is with the intellectual attitudes of the general public, not just with protecting speech activity thought to have important informational value. His own agenda includes changing those attitudes, but he recognizes that courts, and especially the Supreme Court, have the primary educative role. The task of the free speech enterprise was not simply to guard against a regime of seditious libel, but the far more difficult one of creating a kind of democratic personality. The role of the courts in this ongoing process was that of the educator, or teacher. Meiklejohn described the purpose of his own lectures as that of instructing the public about the "basic plan" of American government. To him, the Supreme Court was our foremost "teacher" on that subject: "The Supreme Court, we have said, is and must be one of our most effective teachers. It is, in the last resort, an accredited interpreter to us of our own intentions."[27]

What is being taught is much more grand than when to employ legal coercion against speech. At one level Meiklejohn is concerned that people are too prepared to sacrifice their decision-making authority to the government rather than bear the burdens of intellectual responsibility called for by the idea of self-government.[28] And so the "theory" emphasizes the theme of democracy and self-determination, as part of the original compact of the government, and the act of prohibiting speech is suggestively interpreted either as a display of intellectual cowardice or as an attempt by the government to usurp the sovereignty of the people by depriving them of information deemed "too dangerous" for them to hear. Insistence that the speech—even extremist speech—be tolerated becomes a display of mastery over the "fears" about ideas, for to "be afraid of ideas, any idea, is to be unfit for self-government."[29] It is to be "insecure," "ignorant," and to "flinch." But to tolerate speech is to "face up to ideas," to be "fearless," "unflinching," and otherwise "self-reliant."[30]

The end being sought is not just wise decisions but a more general identity. Meiklejohn offers the possibility of being capable of self-control, a matter always in some doubt and therefore positively established by the insistence that even the most worthless speech never be foreclosed, by oneself or by others. This is the image that would become so attractive to later scholars, who would sense in Meiklejohn's argument the possibility of being an "autonomous" person.[31]

But Meiklejohn's ideas about what tolerance of speech can stand for went well beyond this. He aspired not to an individual self-determination but to a collective one. He speaks in the language of "citizens," the "community," the "general welfare," and the "common good."[32] Within the community he insisted on a particular set of intellectual characteristics. Self-governance for him was not just facing up to every idea, however wrong or dangerous it might be regarded, to be followed by a single count of noses. In particular, it was emphatically not a matter of each citizen's merely figuring out what was "good for him" and then voting on that basis. For Meiklejohn, true self-government was conducted according to a certain intellectual standard by which all citizens think and vote in terms beyond themselves, objectively and in pursuit of the general, collective welfare.[33] Beliefs could be assessed as good or bad, true or false, and part of the capacity we seek is to be better able to render judgments about them. Some beliefs were true, and one of the functions of the First Amendment was to make possible the communication of truths by those who had it to those who did not.[34] Meiklejohn himself gives the impression of a man who believed he possessed the truth about the right kind of community values.

The structure of the "theory" reflects this agenda. Free speech is defined as serving only a "collective interest," and the ideal speech that is deserving of protection is that of the "citizen who is planning for the general welfare." Most important of all is the function of extremes in this affair. In Meiklejohn's terms, the act of toleration, especially of extremist views, becomes a means by which the society both communicates and affirms these commonly shared principles. That the principles are contrary to those of the speaker does not take away from the possibility of communication and affirmation; the conflict instead creates the opportunity and actually enhances it. Just as sociologists could see in the process of *punishing* deviant behavior a method of community self-definition, Meiklejohn could sense the same phenomenon at work in the *toleration* of such behavior, in the special context of free speech. What Meiklejohn proposes as the meaning of tolerance is the demonstration, or proclamation, of confidence in and commitment to one's belief. The community retains a kind of psychological control over the event by virtue of the theory itself, since the speech is tolerated only because it serves the collective interest of the society.

Thus, for Meiklejohn, the First Amendment embodies an intellectual life in the broadest sense, one to which we aspire under the tutelage of the Supreme Court. Self-government is defined to mean not just the absence of governmental censorship but also a set of attitudes about how each member of the community should relate to every other member and about how we should approach the issues that confront us. As for extremist speech, speech that, for example, advocates the destruction of the very system Meiklejohn himself seeks, one can envision at least two functions served by tolerance. The first, which we noted in chapter 4, relates to the pedagogical role the Court assumes under the aegis of free speech. The instructional advantages of extremes are, of course, well known and no doubt did not escape the intuitive attention of an educator like Meiklejohn. The drama and tension of aberrant behavior has, at a minimum, an attention-commanding capacity, which somewhat ironically means that the system can beat the radicals at their own game. But there is a deeper, more substantive gain from toleration.

For it is extremeness itself and the very fears evoked by it that make tolerance of free speech mandatory under a regime like Meiklejohn's. To a society that seeks to develop a certain capacity, especially one of security and control, toleration can help to establish or prove symbolically the arrival of that capacity. Often, the harder something is to do, the more symbolic meaning the doing of it carries. For speech that attacks and challenges community values, the act of toleration serves to both define and reaffirm those values; the act of tolerance implies a contrary belief and demonstrates a confidence and security in the correctness of the community norm. Through toleration, in short, we create the community, define the values of that community, and affirm a commitment to and confidence in those values. To think of freedom of speech in this way evokes Seneca's prescription on mercy: its primary value, when implemented by the ruler, is in what it bespeaks of the party who is merciful, his confidence and security and self-restraint in the face of challenges to that authority:

> Every house that mercy enters she will render peaceful and happy, but in the palace she is more wonderful, in that she is rarer. For what is more remarkable than that he whose anger nothing can withstand, to whose sentence, too heavy though it be, even the victims bow the head, whom, if he is very greatly incensed, no one will venture to gainsay, nay, even to entreat—that this man should lay a restraining hand upon himself, and use his power to better and more peaceful ends when he reflects, "Any one can violate the law to kill, none but I, to save"? A lofty spirit befits a lofty station, and if it does not rise to the level of its station and even stand above it, the other, too, is dragged downward to the ground. Moreover, the peculiar marks of a lofty spirit are mildness and composure, and the lofty disregard of injustice and wrongs. . . . Cruel and inexorable anger is not seemly for a king, for thus he does not rise much above the other man, toward whose own level he descends by being angry at him. But if he grants life, if he grants position to those who have imperiled and deserve to lose them, he does what none but a sovereign may; for one may take the life even of a superior, but not give it ever except to an inferior.[35]

## II

To understand Meiklejohn's argument helps to point up, by contrast, that of Holmes, whose intellectual values, while no less universal in character and no less implicated in his defense of free speech than those of Meiklejohn, were nonetheless strikingly different. Meiklejohn properly sensed the difference, though he seems throughout much of his essay to be confused about the real point of disagreement. Many had attacked Holmes and his principal defender, Chafee, for harboring too great an attraction for free speech. But no one, until Meiklejohn, had really attacked the intellectual framework that underlay Holmes's defense of free speech.

Meiklejohn had difficulty, however, in locating the true conflict between himself and Holmes. He spends too much time denouncing Holmes for the "clear and present danger" test, which Meiklejohn seems to misunderstand. He bridles at the suggestion that whenever speech becomes dangerous, the government can intervene and call a halt to further discussion. Purporting to be an advocate of an "absolute" standard for public expression, Meiklejohn appears to stand for a more rigid level of protection. But the differences are more apparent than real: Meiklejohn distances himself from Holmes only by exaggerating the actual meaning of the "clear and present danger" test, by making no allowance for the shift in Holmes's thought in the years between *Schenck* and *Whitney*, and by failing to acknowledge the flexibility inherent in his own purportedly absolute standard. By the end of the discussion, which begins with a severe castigation of Holmes and his test, Meiklejohn is basically in agreement with the reformulation of the "clear and present danger" test that Brandeis announced in *Whitney* (with, of course, Holmes's concurrence).[36] One is left puzzled as to why Holmes should be regarded as such a villain when the only fault finally attributed to him is that he had earlier announced an overly broad principle, which, though it took him a few years to do so, he did ultimately get right. Even if it were true that "the great majority of [Holmes's] colleagues were taking very seriously the assertion of Mr. Holmes that whenever any utterance creates clear and present danger to the public safety, that utterance may be forbidden and punished,"[37] it seems quite unfair to lay all the blame on Holmes, all the more so when it is conceded that "[n]o one, of course, believes that this is what Mr. Holmes intended . . . to say" and that Holmes "spoke out with insistent passion" as his colleagues misinterpreted and misapplied his unfortunate test.[38]

This nondispute is representative of a good deal of misdirection in free speech argument. As in the reaction to so much of the debate over the utility and meaning of the "clear and present danger" test and the so-called absolute standard, one comes away from these vigorous exchanges puzzled by what the fury is all about. Meiklejohn's tussle with Holmes suggests that what appears on the surface as a disagreement over tests and standards is really only superficial and is actually motivated by deeper and more fundamental disagreements over what basic intellectual character we should be seeking.

A more careful reading of Meiklejohn is therefore required to understand the real and profound differences between him and Holmes. The disagreement centers not on whether to protect certain speech but on the issue of the "moral and intellectual foundations of a self-governing society," as Meiklejohn was ultimately to put it himself.[39] Meiklejohn's commitment to and confidence in community values run squarely into Holmes's insistent proclamations of ultimate relativism in human affairs. The dispute is about basic intellectual values, which inform the recommendation of tolerance in each instance.

Holmes's writings on the First Amendment reflect his well-known skepticism.[40] The clearest and most celebrated of these statements is in his *Abrams* dissent, in which he addressed the "logical" impulse to intolerance and which we must consider once again in our effort to untangle the multiple meanings of these classic texts of the First Amendment:

> Persecution for the expression of opinions seems to me perfectly logical. If you have no doubt of your premises or your power and want a certain result with all your heart, you naturally express your wishes in law and sweep away all opposition. To allow opposition by speech seems to indicate that you think the speech impotent, as when a man says that he has squared the circle, or that you do not care whole-heartedly for the result, or that you doubt either your

power or your premises. But when men have realized that time has upset many fighting faiths, they may come to believe even more than they believe the very foundations of their own conduct that the ultimate good desired is better reached by free trade in ideas—that the best test of truth is the power of the thought to get itself accepted in the competition of the market, and that truth is the only ground upon which their wishes safely can be carried out.[41]

Like so many of Holmes's opinions, his remarks on the nature of intolerance bear a strange mix of bluntness and ambiguity. Read against the background of Holmes's general views, however, the statement from *Abrams* carries a special meaning. While it appears to affirm the possibility of "truth" in human affairs, it does so by foreclosing any meaningful role for the individual other than as a component in a larger marketplace. It instructs us that wanting to believe in the truth of our beliefs is a natural aspect of the human condition, but urges us to overcome that tendency. So many "fighting faiths" have come to nought, it is said, that it is inadvisable to accept the validity of anything we happen to believe at the moment. We had best put our faith in the outcome of the market, perhaps a secular analogue to the comforting illusion of prior centuries that the king spoke with divine authority.

Ultimately, the intellectual posture advocated by Holmes was as encompassing as that of Meiklejohn, though each was fundamentally at odds with the other. For both men the toleration of extreme speech took on special meaning. To Holmes, however, this meaning derived from his view that intolerance is the product of certitude of belief; Holmes shared Emerson's sentiment that "[t]he grossest ignorance does not disgust like this impudent knowingness."[42] In Edmund Wilson's character sketch of Holmes in "Patriotic Gore," he notes of Holmes that the "certainty of one's moral rightness, the absolute confidence in one's system' always set up in him the old antagonism."[43] Wilson provides evidence from Holmes's letters:

"He seems to me," he writes Harold Laski in September, 1918, of the pacifist activities of Bertrand Russell, "in the emotional state

> not unlike that of the abolitionists in former days, which then I shared and now much dislike—as it catches postulates like the influenza"; and in October, 1930, when he has been reading Maurice Hindus's *Humanity Uprooted*, "His account of the Communists shows in the most extreme form what I came to loathe in the abolitionists—knave or a fool. You see the same in some Catholics and some of the 'Drys' apropos of the 18th amendment. I detest a man who knows that he knows."[44]

Holmes "detested" such people perhaps, as Wilson suggests, because he was always reacting against his Calvinist heritage or because of his strong sense of superiority to the common run of people or, which seems most persuasive of all, because of his sense—in part the product of Holmes's experience as a soldier in the Civil War—that belief is a straight road to killing one another. Holmes writes to Laski:

> "Pleasures are ultimates, and in cases of difference between ourself and another there is nothing to do except in unimportant matters to think ill of him and in important ones to kill him. Until you have remade the world I can class as important only those that have an international sanction in war."[45]

Echoing the passage in his *Abrams* dissent on intolerance, he writes:

> "But on their premises it seems to me logical in the Catholic Church to kill heretics and [for] the Puritans to whip Quakers—and I see nothing more wrong in it from ultimate standards than I do in killing Germans when we are at war. When you are thoroughly convinced that you are right—wholeheartedly desire an end—and have no doubt of your power to accomplish it—I see nothing but municipal regulations to interfere with your using your power to accomplish it. The sacredness of human life is a formula that is good only inside a system of law."[46]

Thus, while Meiklejohn sought to construct a view of tolerance as constituting an affirmation of belief, Holmes sought to view it as a commitment to an intellectual stance of self-doubt. For Holmes, the more unbelievable the idea, the more the capacity for self-doubt was tested.

Holmes's intellectual posture during the *Schenck-Abrams* period is nicely portrayed in an exchange of letters (only recently published) between him and Judge Learned Hand, who appears to be in substantial sympathy with Holmes's intellectual outlook.[47] Responding to a letter in which Hand declares that "Tolerance is the twin of Incredulity," Holmes writes that he "agree[s] with it throughout." In the course of the letter Holmes describes himself as being committed intellectually to a course of self-doubt, however unsuccessful the effort to fulfill that commitment: "When I say a thing is true I mean that I can't help believing it—and nothing more. But as I observe that the Cosmos is not always limited by my Cant Helps, I don't bother about absolute truth or even inquire whether there is such a thing, but define the Truth as the system of my limitations."[48]

For his part, Meiklejohn detected the intellectual ethic of relativism in Holmes's claim for free speech and sensed its implications. He criticized Holmes for having "no adequate account of the deeper social ends and ideas upon which the legal procedure depends for life and meaning." He saw Holmes's attitude as "representative . . . of his time and country," a state of mind that Meiklejohn argued was culturally debilitating. Importantly, he saw a connection between the attitudes embodied in the principle of free speech and the attitudes people generally bear. So, he charged that Holmes's relativism had contributed to an excessive "individualism" in the country, and worse, an "intellectual irresponsibility" where "private interest" is given free reign under the rationalization that the "competition of the market" will shed the bad and save the good. It was a license to forgo the burdens of intellectual effort and objective truth for the simpler and attractive path of self-preference.[49]

The conflict between Meiklejohn and Holmes over the meaning of free speech was rooted in a shared effort to develop an intellectual capacity that would be reflected in, and stretched by the act of tolerance toward speech—but for one it was the capacity of shared belief and community, and for the other it was the capacity of self-doubt and individualism. One theory was centripetal, the other centrifugal.

## III

For both Holmes and Meiklejohn, the concept of free speech provided an opportunity to instill or create a particular intellectual outlook. Preserving the activity of speech itself is not the end to be secured, but rather the reformation of perceived undesirable intellectual tendencies and the substitution of other tendencies.

The dispute is replicated elsewhere in the classic free speech literature. This is, for example, the basic division in the argument Mill used in constructing his plea for liberty of speech. His question is this: Should one think of one's beliefs as "true," but nevertheless be tolerant of the expression of untruths, or should one think of them as potentially untrue and proceed to justify tolerance on that basis? Despite the inconsistency of the two premises for a free speech theory, Mill advanced both as alternative grounds. More recently, one finds this tension between competing bases for tolerance reproduced in Bickel's last book, *The Morality of Consent*, where the argument vacillates between a plea for a Burkean acceptance of life without universal truths and a recognition of the destructiveness of relativism and the corresponding need for deeply embedded community values.[50]

Neither notion of tolerance can by itself provide an acceptable or stable foundation for free speech. Let us return to Meiklejohn and Holmes to demonstrate the proposition. Both are doomed to fail because what each offers is unacceptably extreme as a general intellectual posture and because, with respect to the actual functioning of free speech, each is built on an internal contradiction. Each offers a vision that, while psychologically highly important and valuable, is also dangerously overextended.

Consider Meiklejohn. We have observed that he presents an attractive vision of the self-governing community, modeled on an idealized New England town meeting where people carry on under a set of shared values that includes a belief in the capacity for self-control (self-determination) and a belief in the spirit of cooperative striving for the general interest. It would be comforting to know what you believe, to be confident of the bases and reasons for your belief, and to feel in control of your capacity to re-create the environment to correspond to those beliefs. The world we live in, however, and the world Meiklejohn lived in, can itself be undermined by speech. In his essay Meiklejohn argues that it is a function of the First Amendment to let those members of the community who have arrived at a "truth" communicate it to others. But suppose they do so and it is not received by others as a truth at all, but as a falsehood? In a self-governing society, people then face a dilemma when they believe deeply in the truth of some idea, yet others do not appear to share the same feeling.

Meiklejohn himself reveals the fragility of a position of tolerance constituted on a foundation of right belief. At the very end of the essay, he rues the lack of opportunity to apply his theory to concrete examples of the limits of free speech. He finds it "essential, however, to mention one typical failure [of our national education] which, since it has to do with the agencies of communication, falls within the field of our inquiry." The "failure" is "commercial radio." Here was a "new form of communication" that had "opened up before us the possibility that, as a people living a common life under a common agreement, we might communicate with one another freely with regard to the values, the opportunities, the difficulties, the joys and sorrows, the hopes and fears, the plans and purposes, of that common life." Perhaps, he says, "amid all our differences, we might become a community of mutual understanding and of shared interests."[51]

"But never was a human hope more bitterly disappointed." While in the beginning it deserved the protection of the First Amendment, it was not "entitled" to it now. Why not? Because "[i]t is engaged in making money." Radio "is not cultivating those qualities of taste, of reasoned judgment, of integrity, of loyalty, of mutual understanding upon which the enterprise of self-government depends. On the contrary, it is a mighty force for breaking them down. It corrupts both our morals and our intelligence."[52]

Here, in one surprising passage, Meiklejohn reveals the limits of the concept of the "right to hear," of the notion that free speech means we are not "afraid" of dangerous ideas, and that "[t]o be afraid of ideas, any idea, is to be unfit for self-government."[53] In the end, according to Meiklejohn, we cannot afford to tolerate all forms of "subversive" speech, but only those forms that are not really dangerous because the vast majority already dislike them. In the quest for values through enforced toleration, the values can so easily end up taking precedence over the idea of toleration itself. Meiklejohn's "public interest" notions become subtly transmuted into a justification for suppression.

The alternative of self-doubt bears its own seeds of instability. It moves easily into an attitude of superiority or, alternatively (as Meiklejohn said), into that of self-interest and nihilism, both of which can erode social cohesion.[54] Yet its problems as a foundation for free speech are even more fundamental. Perhaps individuals can tenably accept toleration on the conviction that nothing is true, but their actions become much less defensible when they employ the power of the state to inhibit others from being intolerant. In this instance, the individual is demanding an intellectual outlook of self-doubt from those who wish to insist on belief. The question may properly be asked why the belief that "the ultimate good desired is better reached by free trade in ideas" should not be subjected to the same self-critical realization that "time has upset many fighting faiths." If we are better off permitting the "play of forces" and supremacy through "free competition," then why should the same evolutionary process not play itself out in the arena of speech as well?

The attitude reflected in these rhetorical questions can be found in Holmes's early free speech decisions, as in *Schenck*, where tolerance is expressed as the need for judicial acceptance of a nation driven to intolerance by war-generated passionate convictions of true belief.[55] By the time of the *Abrams* case Holmes's wish for self-doubt leads him to insist that others too be self-doubting, and so the free speech principle is enforced. In chapter 1, the question of explaining Holmes's turnabout between *Schenck* and *Abrams* was noted. On that issue I find it a more congenial hypothesis that Holmes's dramatic change of position was not merely (as some have suggested) an unsuccessful strategy employed in *Schenck* to get a good test at the expense of affirming a conviction, or the result of an interim education in the merits of free speech, or the attempt of an isolated man who felt unappreciated to court the favor of liberals,[56] but rather the natural, perhaps inevitable, shifting about on a position premised on the sands of relativism and self-doubt. Relativism, at least in the context where one is called on to resolve disputes between two parties, is a posture ultimately at war with itself.

Clearly, Holmes himself was uncertain about the appropriate implications to be drawn from the acknowledgment of a relativistic universe. In a letter to Hand discussing his thoughts before "the statue of Garrison on Commonwealth Avenue, Boston," Holmes says that if he were "an official person I should say nothing shall induce me to do honor to a man who broke the fundamental condition of social life by bidding the very structure of society perish rather than he not have his way—Expressed in terms of morals, to be sure, but still, his way." As the "son of Garrison," alternatively, he should find himself thinking differently, taking the view that "every great reform has seemed to threaten the structure of society,—but that society has not perished, because man is a social animal, and with every turn falls into a new pattern like the Kaleidoscope." As a philosopher, however, he would believe them to be "[f]ools both, not to see that you are the two blades (conservative and radical) of the shears that cut out the future." But it is the "ironical man in the back of the philosopher's head" who is given the last word, saying of the philosopher that he was the "[g]reatest fool of all" for not seeing that "man's destiny is to fight" and urging him to "take thy place on the one side or the other, if with the added grace of knowing that the Enemy is as good a man as thou, so much the better, but kill him if thou Canst."[57] Holmes himself could only embrace his own skepticism with misgiving and doubt.

## IV

Tolerance is a complex psychological state. Every society, just like every individual, must develop methods of coping with behavior that manifests ways of thinking regarded as objectionable but which, for one reason or another, simply must be tolerated. Some way of thinking must be constructed that will assist those who have to be tolerant in dealing with the impulses and internal doubts about their own identity that are raised by the troublesome behavior. Religious societies may emphasize the responsibility of an independent evil body (like the devil) or stress that the divine being will properly take care of the heretical behavior at an appropriate time (as at the Day of Judgment). Other groups will construct other methods of response. It may help to classify certain behavior as the product of insanity, as the activity of a deranged and an alien mind. The techniques and responses vary greatly, but which techniques and responses are selected will indicate much about the group that selects them. In this society the terminology of "rights" sometimes seems to perform this sort of psychological function of providing an automatic, and socially accepted, way of separating oneself from the acts of others, which is usually an important predicate for tolerance.

But the responses offered by Holmes and Meiklejohn are more sophisticated elements in that process and, at least in Holmes's prescription, also well known to us. We often respond to difficult behavior with the attitude of "Who knows" or "Who's to say." Meiklejohn's recommended attitude, however, is also always powerfully present. "I, and we, know what our beliefs are and it is only doubt that is revealed in intolerance." Denying any legitimate role for belief or affirming belief and using toleration as a demonstration of commitment to belief are two alternative, and contrasting, methods of response.

As we have seen, however, both perspectives have radiating implications for behavior generally, transcending the narrow area of legal coercion against speech. But free speech provides the occasion for such a generalized effect. We can agree with Kalven that at its core the First Amendment prohibits the government from severing a democratic relationship with the citizenry by instituting a regime of seditious libel. And *New York Times Co. v. Sullivan* evoked this kind of First Amendment imagery—even if the reality was quite distant from it. But the securing of that victory would not complete the role to be played by the First Amendment. It would only initiate it, for the idea of free speech, at least as it has come to be known in this century, addresses far broader questions: it can also speak to the intellectual makeup and character of the society.

In that role free speech is a complex enterprise that has a more involved function than preventing governmental interference in the democratic process, maximizing the flow of data, or protecting the rights of speech for minorities against tyrannical majorities. Meiklejohn and Holmes offer competing intellectual bases for free speech. Each proves unacceptably limited. To broaden the inquiry into the role of free speech as a forum for defining certain fundamental intellectual values is to provide a foundation from which we can better evaluate our previous experience with the concept as well as contribute to the development of a more coherent and stable general theory for the future.

Besides their concrete and particular consequences, which sometimes, to be sure, are considerable, free speech cases also constitute fundamentally symbolic social activity. They are like theater, especially classical Greek drama, in which extreme behavior takes place that is socially defined through the context of the litigation stage. Sometimes the value of a given case is to be found more in its dramatic potential than in its particular factual importance. This perspective helps to account for a highly interesting feature of First Amendment jurisprudence: it includes an extraordinary number of disputes that, looked at in isolation, seem nothing short of trivial and socially inconsequential.

One might, for example, reconsider the differences between two major First Amendment cases, *Sullivan* and *Cohen*. Is the primary difference that *Sullivan* defines free speech in terms of seditious libel, offering a systemic vision in which the discussions of the citizens are protected by the courts against governmental hindrance, and *Cohen* defines it more in terms of a speaker's right to self-fulfillment? I think not. The most important differences between those two major opinions are in the intellectual attitudes that each projects, both implicitly and explicitly, especially as they concern the major social conflicts of their times—one the conflict over racial segregation and the other over the Vietnam War.

In *Sullivan* the Court was fundamentally in sympathy with the objectives behind the speech at issue. Indeed, the message of racial equality was one that the Court itself had fostered, and was seeking to enforce, in its desegregation decisions. It could not escape notice that the act here of penalizing this particular speech was a direct challenge to the civil rights movement—an act of symbolic expression, as it were—signifying a determination to continue a policy of resistance. The Court's own will to press ahead was therefore ultimately an issue in the case.

In the face of this broader reality, the Court's rhetorical tact is intriguing. It first offered encouragement for political speakers who wished to promote their beliefs; it encouraged dissent by pronouncing at the outset a "profound national commitment to the principle that debate on public issues should be uninhibited, robust, and wide-open."[58] This same theme was then developed with an apologia for belief's natural tendency toward excess. The Court quoted an earlier decision where it was said of religious and political views that "[t]o persuade others to his own point of view, the pleader, as we know, at times, resorts to exaggeration, to vilification of men who have been, or are, prominent in church or state, and even to false statement."[59]

The opinion advances this message of determination in the face of provocation by the manner in which it defines the act of toleration generally. Toward the end of the Court's analysis it describes the Court's own posture of self-restraint in the face of harsh criticism. The "concern for the dignity and reputation of the courts," the Court says, referring to an earlier case, "does not justify the punishment as criminal contempt of criticism of the judge or his decision," even if the "utterance contains 'half-truths' and 'misinformation.' "[60] This is because judges "are to be treated as 'men of fortitude, able to thrive in a hardy climate. . . .' "[61] Here the idea of tolerance is not that speech is a rich source of potential ideas or that criticism might be helpful in arriving at the correct result, but rather that speech is something to be tolerated with "fortitude." The overtones are again ones of firmness of conviction, and they converge with the earlier sympathetic explanation in the opinion about the particular speech challenged in the case. By this single characterization of the posture of tolerance, the Court at once affirms its own resolution to proceed despite acts of provocation and expressions of hostility, like that of the libel judgment it was then reversing, and aggressively ridicules its opponents by impliedly describing their actions as the product of weakness, of lack of "fortitude." As with Meiklejohn's essay, *Sullivan*'s power derives from its definition of toleration as an act involving the affirmation of belief.

The Court in *Cohen*[62] also explored how speech activity ought to be regarded, though there the context was the more traditional one in which the Court itself disagreed with the particular exercise of speech at issue. The opening sentence of the opinion unambiguously stated that though "[t]his case may seem at first blush too inconsequential to find its way into our books, . . . the issue it presents is of no small constitutional significance."[63] This intimation that there had been an abuse of a right was gradually made plain, first as Harlan refrained from repeating Cohen's slogan, then as he heaped scorn through elegant paraphrase ("the inutility or immorality of the draft"),[64] and finally, more explicitly, when he labeled the speech a "distasteful mode of expression,"[65] "a crude form of protest,"[66] and a "trifling and annoying instance of individual distasteful abuse of privilege."[67]

After the speech activity was thus characterized, the question then became how to vindicate its toleration. As with *Sullivan*, the messages were mixed. On the one hand, it was said that this seeming "verbal cacophony is . . . not a sign of weakness but of strength."[68] On the other hand, the final judgment was more skeptical than anything encountered in *Sullivan*: The speech could not be excised because "while the particular four-letter word being litigated here is perhaps more distasteful than most others of its genre, it is nevertheless often true that one man's vulgarity is another's lyric."[69] *Cohen* arose, in the end, more from Holmes's intellectual perspective. The activity was distasteful, but Harlan was reluctant to coerce on the basis of his aesthetic predilections.

## V

We find, therefore, in Holmes and Meiklejohn a recapitulation of the general tolerance function of free speech. Meiklejohn and Holmes were responding not to some single-minded urge for information but to the perceived need to correct what might be thought of as a bias or a deficiency in the way people think about issues when engaged in the social process. Such a bias could be found in social decision making about *every* issue, whether it concerns the limits of speech or nonspeech conduct. But it is the area of speech that has been seized as presenting the opportunity to identify and address that deficiency. Because under either Meiklejohn's or Holmes's view this deficiency in social discourse manifests itself in a tendency to excessive intolerance of speech, speech can be treated as a portion of behavior that is excluded from the social decision-making process. And toleration of speech can be pushed to an extreme, as a kind of opportunity to point up and address what is, in fact, a broader problem. Free speech becomes a concept under which an area of social organization is effectively carved out and rendered immune, not for its uniquely important role in assisting decision making, but as a way of pressing for an intellectual capacity or perspective that is also thought to be of direct value elsewhere.

It is important to recognize the significance of the change in orientation involved in this way of looking at free speech. The writings of Holmes and Meiklejohn are representative of an extremely important, but generally unnoticed, shift in perspective on the meaning and function of the free speech principle. No longer is the premise of the discussion about free speech the assumed need to secure "our" liberty to speak freely against some "other" antagonistic political body or power. The center of focus has shifted away from the purpose of establishing an unrestricted zone of liberty in which we can fulfill "our" interests to speak and to hear the information and ideas we value—what Isaiah Berlin referred to as the idea of "negative liberty"[70]—toward the purpose of creating a zone of unregulated behavior against which "we" can test and develop a general character of mind through confrontation with that behavior—what may be thought of as the goal of exploring the virtue of tolerance. In a sense this latter ambition, or aspiration, can itself be thought of as a variation on that ambiguous concept "liberty"; indeed, it seems to resemble what Berlin classified as the notion of "positive liberty," that is, the "wish on the part of the individual to be his own master." According to Berlin, the mastery sought may be over "external forces" or over irrational passions and impulses internal to the individual or the community.[71] In envisioning these two divergent components of the concept of liberty, Berlin worried about the positive conception's being turned into a source of oppression, a concern to which the discussion here lends weight. What he did not see, however, is how the two conceptions of negative and positive freedom could be yoked together, the former in service to the latter; how the establishment of a zone or field of negative freedom could become a source or method of striving for some sense of positive liberty.

Thus, free speech is a social context in which basic intellectual values are developed and articulated, where assumptions about undesirable intellectual traits are offered and remedies proffered. Both Meiklejohn and Holmes spoke to these issues. Though I have argued that both ultimately failed in their efforts to define a set of intellectual values for society through free speech, the failure in each case was only partial and due to the universality each sought for his own paradigm. Postures of self-doubt and affirmation of belief must be regarded as unstable; each attempts to resolve issues that are ultimately not resolvable, to settle for once and for all what cannot be. Neither intellectual posture alone is a guarantee of desirable social behavior; either, in fact, may become a pretext for giving into base impulses (as the program of the National Socialist Party in Germany was for the violent propensities of the youth who became Brownshirts.)

We are, and must inevitably remain, fundamentally ambivalent toward the process we describe as belief. While Holmes may have been right that belief lies at the core of the impulse to excessive intolerance, it also forms the basis of our system of morality, and we make wide use of it in structuring social behavior. A serious risk of free speech is that the very limited context in which the matter of excessive intolerance is addressed will lead those involved, especially judges, to propose too strong an antidote for the general disease, one people are simply incapable of living with, even in the realm of free speech itself. That was the fundamental error of both Holmes and Meiklejohn.

# 6

# Drawing Lines and the Virtues of Ambiguity

From the outset, it was indicated that this was not going to be a book about free speech doctrine but, rather, one about a new function the free speech principle has assumed in American society. I intend to abide by that limitation, although it ought to be understood, too, that the limitation is not just self-imposed. Free speech has become too large an area of the law for any single book to trace the implications of a theory.

Nevertheless, it seems desirable to give at least some indication of the kind of implications the tolerance function of free speech might have for shaping the principle's contours. In this chapter, therefore, we take a look at several areas of the basic structure of First Amendment doctrine and consider them afresh in the light of the general tolerance theory. In chapter 7 we take up the subject of potential biases that are internal to the process of implementing the free speech principle itself.

## I

Let us begin with the basic First Amendment problem of deciding how to think about how far we ought to go in letting people advocate action or behavior that threatens to bring about adverse consequences to deeply held social values, such as the preservation of the democratic system of government or the maintenance of lawful processes of social decision making or the adoption of various policies widely regarded as immoral—for example, genocide or the oppression of various minority groups within the society. To what extent should these speech activities be tolerated within the society, at least under the principle of freedom of speech? What test is appropriate for deciding upon those limits?

In chapter 2 we began discussing the complex ways in which speech—more accurately, imposing limits on a society's use of legal restrictions against speech—might bring harm to the society. We sought a more complete understanding of the range of potential costs inherent to a free speech principle than the classical model seemed to recognize. Chapter 4 offered a theory for why the principle of free speech might be pushed to what seems like an extreme, a theory that would explain why an even more complete accounting of the risks and harms would not lead us to draw the limits of free speech at an earlier point along the spectrum of harm from expression. But to say that we might sensibly choose to pursue the free speech idea to an extreme is not to say that we ought to tolerate speech under all circumstances and conditions. It would be senseless to do so. Free speech should not and can not mean that we will suffer any and all consequences that speech may bring. Virtually none of us would presumably press for the right of an individual to reveal information in a wartime setting that, if known by the enemy, would almost certainly lead to the country's defeat and the loss of democratic freedoms. The First Amendment, no less than the Constitution as a whole, should not, as Justice Robert Jackson warned years ago, be turned into a "suicide pact."[1]

Of course, this is precisely why, despite all the rhetoric that free speech is an "absolute," the doctrinal structure of free speech has always contained a number of "exceptions" for regulation of speech activity. For Holmes, in *Schenck* this was the point at which the society faced a "clear and present danger that [the words] will bring about the substantive evils that Congress has a right to prevent." It was, he said, "a question of proximity and degree."[2] In *Abrams*, Holmes seemed to fortify the test by emphasizing the elements of imminence and magnitude of the potential harm: "I think that we should be eternally vigilant against attempts to check the expression of opinions that we loathe and think to be fraught with death, unless they so imminently threaten immediate interference with the lawful and pressing purposes of the law that an immediate check is required to save the country."[3] Subsequently, in the *Whitney* case, Brandeis restated the formula: "But even advocacy of violation [of law]," he said, "however reprehensible morally, is not a justification for denying free speech where the advocacy falls short of incitement and there is nothing to indicate that the advocacy would be immediately acted on."[4] The danger, he added, had to be "serious": "Moreover, even imminent danger cannot justify resort to prohibition of these functions essential to effective democracy, unless the evil apprehended is relatively serious. . . . The fact that speech is likely to result in some violence or in destruction of property is not enough to justify its suppression. There must be the probability of serious injury to the State."[5]

With these revisions—or perhaps only clarifications—of the "clear and present danger" test, Meiklejohn found he could agree with the limits on speech imposed under it. Though Meiklejohn thought the original "clear and present danger" test "radically altered," and even "abandoned," by the "modifications" that both Holmes and Brandeis had originally introduced, he himself agreed that free speech could be suspended or limited in an "emergency," when "no advocate of the freedom of speech, however ardent, could deny the right and the duty of the government to declare that public discussion must be, not by one party alone, but by all parties alike, stopped until the order necessary for fruitful discussion has been restored."[6] The conditions for declaring such an "emergency" were "to be found, not in the dangerous character of a specific set of ideas, but in the social situation which, for the time, renders the community incapable of the reasonable consideration of the issues of policy which confront it. In an emergency, as so defined, there can be no assurance that partisan ideas will be given by the citizens a fair and intelligent hearing. There can be no assurance that all ideas will be fairly and adequately presented."[7]

Given his reputation as one of the staunchest proponents of taking free speech as an "absolute," Meiklejohn's acceptance of the necessity to limit speech in certain social circumstances reflects the law's persistent willingness to recognize some such limit. Variations abound. One prominent alternative to the "clear and present danger" approach is that usually referred to as the "incitement" test, which Learned Hand proposed in a World War I case known as *The Masses*,[8] when he was sitting as a federal district court judge. Hand's idea was that explicit advocacy of illegal action could be proscribed, whatever the actual risks of successful persuasion, but that free speech would protect a speaker who "stops short of urging upon others that it is their duty or their interest to resist the law."[9] The specificity of the line, Hand hoped, would yield consistency in enforcement, but this was offset by the "Mark Anthony" problem, that is, that under Hand's test the society could not restrict *implicit* appeals for lawlessness, however great their danger.[10] Some people have expressed the view that present First Amendment case law embodies a test that combines the "clear and present danger" test of Holmes and Brandeis with the incitement test of Hand; that conclusion was reached upon an interpretation of a statement from the Court's 1969 decision in *Brandenburg v. Ohio*: "[T]he constitutional guarantees of free speech and free press do not permit a State to forbid or proscribe advocacy of the use of force or of law violation except where such advocacy is directed to inciting or producing imminent lawless action and is likely to incite or produce such action."[11] Whether "advocacy" and "inciting or producing" were intended to refer only to explicit appeals to lawlessness, however, is not self-evident.

Initially, what is significant is not the particular test to be ultimately settled on but rather the fact that we begin with a common agreement that some limit must be established for free speech—that we will not insist that the society exercise self-restraint toward all speech, however dangerous. At the same time, it is equally important to see that the tests of the "clear and present danger" genre appear to focus on a very narrow range of potential social harm from speech—that is, on the risk that the audience will be persuaded to turn from talk to action of a kind deemed especially harmful to the society. Based on the discussion in chapter 2, it should be clear that this is not the only way in which tolerating speech can bring injury to a community. It would be surprising, therefore, to find this the only permissible ground for regulating expression. In fact, over time several "categories" of speech have been more or less excluded from protection under the First Amendment, even though they pose little in the way of a threat of immediate persuasion. These categories are, primarily, fighting words, libel, and obscenity.

In 1942, in the case of *Chaplinsky v. New Hampshire*, the Supreme Court first identified these areas as excluded from protection under the First Amendment. "[S]uch utterances," the Court reasoned, "are no essential part of any exposition of ideas, and are of such slight social value as a step to truth that any benefit that may be derived from them is clearly outweighed by the social interest in order and morality."[12] Many have objected to the "two level" kind of analysis *Chaplinsky* seemed to propose, under which certain areas of speech activity could be simply disregarded for purposes of First Amendment analysis.[13] Subsequently, of course, beginning with *New York Times Co. v. Sullivan*, the libel area was substantially reshaped to better fit it to First Amendment considerations. In a series of cases initiated by *Sullivan*, the Court developed a pattern of constitutional rules by which it is generally held that, at least with respect to public officials and public figures, a state may not award damages for injury to an individual's reputation arising out of misstatements of fact unless the statements were made with what is called "actual malice," that is, with knowledge of falsity or with reckless disregard of their falsity. The field of state regulation left untouched by libel is still very great, however, and the basic rationale offered by *Chaplinsky* for permitting this regulation has resounded through the cases ever since: In *Gertz v. Welch,* for example, the Court said that libelous statements were not entirely protected because '[n]either the intentional lie nor the careless error materially advances society's interest in 'uninhibited, robust, and wide-open' debate on public issues" and because "there is no constitutional value in false statements of fact."[14]

In the obscenity field, comparable efforts have been exerted to impose some First Amendment limits on regulation; though often the case for regulation of pornography is put on the ground that it will "persuade" people to commit crime, generally the rationale offered has to do with the minimal "social value" of such expression:

> The question about obscenity is not whether books get girls pregnant, or sexy or violent movies turn men into crime. To view it in this way is to try to shoehorn the obscenity problem into the clear-and-present-danger analysis, and the fit is a bad one. Books, let us assume, do not get girls pregnant; at any rate, there are plenty of other efficient causes of pregnancy, as of crime.[15]

The constitutional test for obscenity, which was put forward in the beginning cases in the area, did not ask whether audience misbehavior was imminent, but whether the work contained anything of "redeeming social value."[16]

So, too, the exception for "fighting words" has survived a series of different courts and cases testing its limits, though now it is commonly said to be narrowly defined by the case law—as may be recalled from the opening discussion of its role in the *Skokie* controversy—to apply only to personally insulting remarks made in face-to-face encounters. The problem of violence from an audience that is "hostile" to the speaker's messages, which is really just an expanded variation of the fighting-words defense of regulating speech, on the other hand, has been rather firmly rejected, with a few exceptions like the Court's 1952 decision in *Beauharnais v. Illinois*,[17] which involved the constitutionality of an Illinois group libel statute. The fighting-words exception, therefore, has been kept rather penned in, though its retention has continued over time to be justified primarily on the basis that such speech possesses little if any "social value."[18]

Today it is common to treat these various areas of speech activity (that is, fighting words, libel, and obscenity and some version of clear and present danger) as constituting the central exceptions to the First Amendment; a method of analysis for First Amendment cases is frequently proposed and applied, in which it is said that unless a regulation of expression falls within one of these categories, it is unconstitutional.[19] In other words, if the state seeks to control or restrict communication in the context of public discussion on the grounds that the speech is "dangerous" or "offensive," in order for it to succeed under the First Amendment it must be able to fit the character of the speech and the circumstances under which the speech is delivered into one of these "categories."

Now, this method of thinking about the limits of free speech does not adequately reflect what ought to be the guiding principles for the necessary task of deciding how much, or what kinds of, speech activity the society must tolerate within its midst. It is too narrow to provide a coherent rationale or sensible criteria by which we can undertake to decide what exceptions ought to be created.

The problem we encounter in developing a rationale for imposing limits on the First Amendment is well illustrated by, and really begins with, the unsatisfactory way in which *Chaplinsky* and its progeny have reasoned their way to the exceptions. Certain speech, the Court said, lacks "social value" and constitutes only a "small step in the search for truth." Because this speech possesses such small benefit for truth seeking, it becomes appropriate to withhold protection, especially when the competing social interests at stake—maintaining "order and morality" and personal reputation and the like—are so significant by comparison. But this way of looking at free speech is fundamentally misguided. For it is not the absence of social value that determines whether the principle of free speech is applicable; indeed, the perceived absence of value is, if anything, a major reason for protection, or, more accurately, for toleration (though the problem here may also lie in there being different levels of meaning in the term *value*). It is self-restraint toward what we believe to be without social value, or, in Holmes's words, what we "loathe and think to be fraught with death," that alchemizes the event into something of real value—not, it may be worth emphasizing once again, because we thereby gain the added measure of guaranteed protection for the speech we do value but rather for the insights and lessons we obtain about ourselves and for the increase in our capacity for toleration generally. This, too, is a "social interest" that must be furthered, though unlike the social interests referred to in cases like *Chaplinsky*, it calls for applying the free speech principle rather than withholding it.

Again, this is not to say that toleration of, or self-restraint toward, *all* speech is mandated under the broader objectives of the First Amendment. Nevertheless, to see the function of free speech in terms of enhancing the capacity for general tolerance is to begin to see where and how limits might be placed. The starting point would seem to be this: Certain extraordinary times and conditions exist in any society in which it is quite simply too much to expect of people that they be self-restrained toward speech behavior, and under which it would be counterproductive to the aspirational aims of free speech to insist on toleration. This starting point for thinking about the limits of free speech would seem preferable to asking whether the expression possesses "social value." In every field of law, and it is no less true in the area of the First Amendment, we must decide as a fundamental question to what extent we shall take ourselves as we are and, on that basis, construct legal principles, and to what extent we shall aim at pushing ourselves to higher levels of behavior. As a general rule, free speech ought to be viewed as being in the reformist camp of legal principles. But there may come a time when that objective can no longer realistically be entertained, when its ends would even be threatened by our pursuing its extension.

The context of fighting words provides a useful example of precisely this type of tension between the desire, through the principle of free speech, to test the limits of self-restraint in the face of disturbing speech behavior and the practical difficulties encountered in applying the principle in a context where self-restraint is, at least for most people, just too difficult to maintain. In the case of highly insulting language, it is surely reasonable to think that little is gained by insisting on self-restraint as a means of learning something about the importance of not succumbing to the types of pressures one feels from challenges of this kind. For centuries, dueling was a common, and at times even socially expected, response to verbal challenges to an individual's personal "honor"—challenges, of course, that evoke the very same issues about the identity of the person verbally attacked that Holmes spoke of in *Abrams* as being at the root of the impulse to intolerance. Today we choose to outlaw such responses, but the residual behavioral needs underlying them may well be not yet sufficiently quiescent that we can fruitfully afford to create an environment in which extreme versions of challenging behavior are legally permitted and violent responses forbidden.

On the other hand, we face a difficult problem of containing such an exception, as many quickly realized, for, if given a loose rein, it could ride roughshod over the entire purpose of free speech. "Hecklers" would be afforded a virtual "veto" over speech activity, and to avoid this means of gutting the principle, we must reject even a high probability of audience violence as a legitimate ground for setting aside the general rule of toleration.[20] The reason for maintaining the principle under such conditions is not just that the anticipated or actual response of violence will interfere with valuable speech activity; rather, it is that containing that response, and learning something about its nature, is in fact a central lesson of the free speech principle.

A similar appreciation of the limited social usefulness of insisting on tolerance, and not of informational and ideational emptiness, can be seen as underlying the exception traditionally granted under the First Amendment for the regulation of pornography. To claim that pornography lacks social value is to put the discussion about the benefits and harms of social tolerance of it on as oversimplified a plane as it does to argue that obscenity causes criminal behavior by inducing audiences to go out and commit sexual crimes. It is only a pretense to think that apart from the risk of immediate persuasion, the society suffers no cognizable injury if it is forced to tolerate, legally, the distribution of pornography. The nature and extent of social harm arising from this sort of speech activity is certainly far more complicated. In the deliberation about whether to create an exception for the regulation of pornography, the relevant inquiry ought to be into the extent to which the attractions of pornography rest in a pattern of thinking that, if allowed to be entertained, will affect people's behavior in many different ways, well beyond the power of legal sanctions to prohibit or control.[21] Attitudes toward family, community, aggression, and so on, may all be linked to the fantasies entertained through pornography. Moreover, the perceived connection between the way of thinking encouraged by pornography and other undesirable social behavior may make pornography an important locus for the community to make a symbolic statement about its general values.

All these considerations have recently come to the fore in the never-ending obscenity debate as a result of the effort by a substantial segment of the women's movement to ban pornographic material because of its perceived linkage to aggression and discrimination against women generally.[22] Whether obscenity "causes" these attitudes is not entirely the point, for such material is claimed to involve a graphic and extreme manifestation of such attitudes and is therefore an appropriate area for the society to symbolically reject, through legal prohibition, such ways of thinking. (In the race area we have a similar recognition of the broad harmfulness of tolerance of speech and of the potential for symbolic statement as certain radio and television productions like "Amos 'n' Andy," children's books like *Little Black Sambo*, and restaurant names like Sambo's have been charged with reflecting, if not necessarily constituting a principal cause of, the pervasive prejudice toward blacks.)

With obscenity, however, something must still be said to distinguish the problematic character of that kind of speech from other kinds, which are also difficult to tolerate and offer important opportunities for symbolic assertions of community values. The answer may rest in the suggestion of psychiatrist Willard Gaylin, who some years ago observed that the problem people have with pornography is in its attraction, or in the fear of its attraction.[23] Our disgust, it seems, is not unrelated to our desires. The real social difficulty posed by obscene material, in other words, may lie in the potential for confusion about what toleration would mean. Obviously, this is a most difficult judgment to make, but because we live in the post-Freudian age, those in the law cannot casually dismiss the claim either that sexual instincts are not easily estimated or that they lie at the core of an individual's, and presumably a community of individuals', identity. The centrality of the sexual instincts to the personality may provide the best explanation we have for the desire to isolate obscenity from the general area of toleration required by the First Amendment. (It should be emphasized, however, that here, as before, I mean to indicate only the direction our inquiry into the issue should take and not its resolution.)

Libel, on the other hand, is a more difficult exception to assess. Once again, it is hard to explain the exception by looking at the lack of "social value" to be derived from false statements of fact injurious to individual reputations, if we compare the relative value of such statements to the truth-seeking enterprise with other speech we insist must be tolerated. As with obscenity, the fact that libelous statements have historically been regulated may count for something, but certainly not everything, given, at the very least, that other areas of speech activity long subject to regulation have been brought within the First Amendment. Perhaps it is the case, as some have suggested, that this society's interest in preserving individual honor and reputation is just that much more deep-rooted than are other interests with which speech regularly interferes.[24]

We may, however, raise one important consideration in the matter of whether to make an exception for libel. It arises from the fact that a single individual has been harmed by the speech act. When one person tends to bear the major brunt of the harm of speech activity, rather than the larger community, the central purposes of the free speech principle are not so likely to be realized by the insistence on toleration. For the tolerance function of free speech focuses primarily on the reform of those who possess *social power*, on the community as a whole or on those who because of their numbers or position effectively hold (or have a reasonable chance of holding) the reins of authority. This is the primary audience that people like Holmes and Meiklejohn, as well as so many others, were addressing as they tried to identify the intellectual character that ought to guide social interaction. Now, in the libel case, the community is actually implementing a system of coercion and punishment (unofficial, of course, but nonetheless, as we have noted before, of substantial power) against the defamed plaintiff. To then insist on toleration of the speech act as a means of mastering restraint is, to say the least, anomalous; this is especially so when, as sometimes occurs with defamatory accusations, the community's coercive response is itself excessive and when, apart from legal adjudication, there is small likelihood that the community response will be ameliorated by any other means ("the truth rarely catches up with the lie," it is frequently said of defamatory statements).

Thinking about the three exceptions—fighting words, obscenity, and libel—in the light of an understanding of the tolerance function of free speech helps us to think more generally about imposing limits on the principle. Specifically, we are led to recognize that the tolerance function must occasionally give way to the reality of human needs. The considerations that lie behind the three well-established exceptions are not isolated to those particular categories, and it would be counterproductive from a free speech standpoint to proceed on any different assumption.

As indicated earlier, we ought to take as a starting point that with some matters, and at some times in our social life, it may be simply too much to expect of people that they tolerate certain speech acts. The adverse social consequences that would be suffered as a result of a blanket, unbending insistence on toleration outweigh the aims of the free speech enterprise. Consider in this regard an example not usually thought to support this judgment about the importance of putting limits on our efforts to push the capacity for toleration, namely, the way religious belief is handled in this society.

The First Amendment, of course, provides not only for freedom of speech and press but also, in another clause, for the guarantee of religious freedom. This, however, is a guarantee both for individuals to practice and advocate their religion freely—which makes it parallel to the freedoms afforded speech generally by the free speech clause—*and* for the insulation of the religious sphere from any official "establishment."[25] This latter proviso is of major interest and importance for present purposes, because its effect is essentially to remove religion from our political life—not entirely, to be sure, but enough so to make it remarkable. It does this by formally divesting policymakers of the power to implement or aid religious programs and also by generally reinforcing what can only be described as an ethic within the society to the general effect that religion and politics are matters that ought to be kept separate. To be sure, the division of religious life and political life into separate spheres is neither rigidly maintained nor consistently advocated; religious clergy not infrequently speak out on public issues and even occasionally hold political office, as occurred during the Vietnam and civil rights era, when religious leaders brought an especially strong moral voice to the public debate, and, more recently, with respect to the abortion controversy. At most points in recent history, however, the norm and the emphasis have been in the other direction: religion is regarded as part of the "private" realm and politics part of the discrete, secular world. When religion seems to be infecting public debate, either because of the strong religious identity of the participants or because of the religious language invoked by those participants, there will typically be, at the very least, expressions of discomfort, of some important line having been crossed.[26]

Why has the society followed this path of maintaining such a generally narrow role for religious thinking, of isolating it from the public sector? A large part of the answer, I think, lies in the tremendous potential of religious belief to produce divisive, even explosive, intolerance.[27] The difficulty with religion, from the standpoint of maintaining a basically peaceful society, is that it is not something people find it easy to talk about. Religious discussions, as every family knows, have a tendency to get out of hand, to bring the discussants to heated interchange, even blows. As we look around the world at the major conflicts of the present time (in Ireland, India, and the Middle East, for example), we find that most are rooted in disagreements over religious belief. In the United States, this potential of religious belief to produce social conflict has been handled by essentially removing religion from public discourse, by turning it into a nonpublic issue.

What is true of religion, people's inability to tolerate open debate and self-doubt over religious beliefs, is also true of other aspects of their lives. We do not expect parents to be "objective" in the way they think about their children, or spouses, or even friends, about each other, at least not nearly to the degree that we expect people to think objectively about how to allocate tax burdens. The way we think about people for whom we feel love or affection is to some considerable extent inconsistent with a frame of mind in which we are continually prepared to reexamine that thinking with cool detachment. Love, like religious faith, requires something of a withholding of critical judgment.

This seems equally true in certain periods of a nation's life. There are times when uncritical belief is vital, when little self-doubt can be tolerated, when wholehearted commitment is needed. The most obvious time in which this is true for a nation is when it is engaged in a full-scale war, especially one perceived as necessary for its survival. Then, the consequences of the messages of tolerance, which Holmes identified for us in *Abrams*, and which now involve a party external to the society, become exponentially more severe. At such times the nation becomes like a family, and it is a matter of personal honor to commit oneself to some greater degree than usual uncritically to the community's cause. I exaggerate, but only to convey a full sense of the thinking likely to be present at such times. One need only look at the conflicts created by dissent in societies recently engaged in war—Britain in the Falklands and Israel in Lebanon[28]—to see the divisiveness and the anger stimulated by continual dissent (though I use these not as examples of appropriate occasions for putting limits on speech but as illustrations of the kind of tensions and internal conflicts that must be taken into account in deciding *when* to tolerate limits).

It is precisely this plea for a sensitivity to the capacities, or incapacities, of communities to handle the insecurities and doubts attendant to toleration that seems to have been at the root of what Holmes was trying to say in his first free speech opinion in *Schenck*, and even later in his turnabout in *Abrams*. Holmes was certainly right in thinking that the "character of every act depends upon the circumstances in which it is done" and that when "a nation is at war many things that might be said in time of peace are such a hindrance to its effort that their utterance will not be endured so long as men fight. . . ."[29] Free speech is and must be an institution with a sensitivity to context, and wartime is one such context in which the capacities for toleration of dissent are notoriously more limited, in which the capacity to tolerate the risks of tolerance typically are significantly reduced, and in which the need to demonstrate commitment and resolve to the cause at hand are correspondingly increased. These concerns underlying the need for intolerance are far greater and more complex than the concern of simply the risk of immediate persuasion. It is the recognition of this larger reality that is suggestively stated in Holmes's famous example in *Schenck* that the "most stringent protection of free speech would not protect a man in falsely shouting fire in a theatre and causing a panic."

The "shouting fire in a crowded theatre" example has always proved strangely Delphic for free speech theorists. It is regularly invoked by anyone seeking to draw some limit on free speech, but the quotation is usually only blindly put forward for the unhelpfully blunt and properly uncontroversial claim that there must be *some* limits to free speech, as the example forcefully demonstrates, while it is, in return, commonly dismissed angrily by free speech proponents as entirely inapposite to controls on regular public debate. Holmes's example has seemed something of an embarrassment, a lapse in recognition of the relevant and the irrelevant, a reflection of an early insensitivity to free speech. Holmes has been accused of offering an "inapt" analogy and a "sterile example" for thinking about the limits of public discourse.[30]

Indeed, there does appear to be a startling irrelevance to the hypothetical. *Schenck*, like other cases of its time and genre, involved political advocacy, albeit of an inflammatory variety. Political ideology is under attack in those cases, not false statements of fact about matters having little or no connection to political or social issues. No one has ever seriously contended that the law of perjury imposes an unconstitutional abridgment of freedom of speech. Furthermore, the context of the hypothetical seems skewed. Whereas the asserted risk with the speech in *Schenck* arose from the expression of an opinion (urging resistance to the draft), the problem in the theater example arises from the general statement of a falsehood. As an illustration, therefore, the example does seem, while obviously correct, both "inapt" and "sterile."

But there is a sense, a very important sense, in which the hypothetical is acutely relevant. The social problem it suggestively raises is the impact of speech on the reactions of the listeners, not only toward the speakers (whether favorable or hostile) but toward each other. The concern here is with the degeneration of relations within the general community. When the speaker in the crowded theater turns on the others, "panic" sets in and the cohesion that formerly kept the group internally civil has now dissipated. The unity and tolerance that formerly held the group together have been replaced by murderous competition and self-interest. Speech, the point is brought home to us, can destroy the collective bonds that normally hold society together.

This is no less true of speech that conveys more abstract ideas than it is of false messages of alarm. Especially in times of crisis, differences and divergences within the population increase exponentially; new ones are created and those that existed before often become more pronounced. But the problem exists in degrees. It was, in fact, the basis of the claim in *Skokie*, noted in chapter 2, that one of the risks of the Nazi march was the disruption of the peaceful coexistence that had tentatively been maintained in the community between Jews, Christians, and blacks. That claim, whatever its ultimate merit or impact on the overall decision in the case, at the very least should have been allowed to be developed and should have been seriously considered. As it was, and still is today, such an argument has no recognized place in the doctrinal structure (except through isolated categories like fighting words) and tends to be met with incredulity or a blank stare instead of acknowledged as a legitimate avenue of inquiry.

What the foregoing discussion amounts to is a plea to recognize the critical importance to society of being at liberty to control or regulate speech activity under certain very limited circumstances or conditions. Whatever one thinks of the validity or the appropriate containment of the traditional exceptions for fighting words, obscenity, and libel, they do (properly understood) at least illustrate or reflect the relevant considerations for exceptions. Perhaps it is odd, but it remains nonetheless true, that by increasing our understanding of the extraordinary power of the impulse to intolerance, and of the concomitant social need to establish some controls over it, we simultaneously become more sensitive to both the inevitability and, indeed, the desirability of that impulse, and likewise to the concomitant need to have that sensitivity reflected in our evaluations of free speech problems. Considered from this angle, it is hardly surprising either that the First Amendment doctrine has over the years had difficulty devising a linguistically fixed test for establishing the degree of toleration required under the principle or that a number of categories of speech have been carved out for more generous levels of social regulation.

The central organizing principle for the area, it would seem, must be this: Whatever verbal formulation is ultimately used as a starting point for free speech analysis, it must be flexible enough to permit, and perhaps even invite, consideration of the wide variety of social harm speech may cause, while also strong enough to reflect the important institutional role of free speech, that the central purpose of the enterprise is to push the boundary of toleration far beyond what would be considered normal by the usual standards of the society. It would be most unwise to insist on toleration, no matter what the social conditions at stake, in every circumstance except that where a specific speaker was about to induce a specific audience to take immediate and serious nonspeech action. The capacity of any society, just as of any individual, for toleration must be expected to possibly vary as conditions within that society change. Given this critical temporal limitation to the relevance of what is done in one time for that in any other, it would seem advisable to prefer an abstract—indeed, one might even say, a conscientiously ambiguous—doctrinal standard. For this purpose the "clear and present danger" standard, which Holmes first proposed, seems a perfectly appropriate formulation, provided we give the term *danger* the wide and sensitive compass it ought to have.

It must be acknowledged, however, that this view of an appropriate standard cuts sharply against a major (some would say, the prevailing) attitude toward how the free speech principle should be applied. Much of the modern literature that deals with the problem of setting boundaries for the First Amendment reflects the attitudes of the fortress model, and most authors incline toward some standard that will provide as little flexibility, or as much certainty, in the application of the principle as possible. The underlying premises of this are several and, by now, self-evident: There will be periods, for whatever causes, of intense social intolerance, when much or nearly all of the population will press for abandonment of the free speech principle; judges are susceptible to public pressure of this kind and will, accordingly, feel inclined to let the majority have its way (as well as themselves sometimes be prone to excessive intolerance toward speech); and the best security we have against a judicial collapse in the face of these pressures is an unambiguous standard and a set of precedents that virtually foreclose any choice but enforcement. Understandably, if you embrace this underlying perspective, any litigation in which communities or the federal government seek to defend censorship, as occurred in *Skokie*, will be viewed with deep alarm, and this explains why judges are typically advised that the best thing they can do for the free speech principle is to dismiss challenges to the principle as expeditiously as possible. Otherwise, it is thought, the delays and costs of litigation alone will be tantamount to censorship. Thus, this is why many free speech proponents in *Skokie* persistently proclaimed it an "easy case," one that ought to have been summarily dismissed and certainly not extended over a period of more than a year, as, in fact, occurred.

While this long-standing and pervasive way of thinking about the structure of free speech doctrine may well be entirely natural (just how natural we will take up in chapter 7), it poses grave obstacles for an effective free speech principle and ought to be resisted.

An initial question, of course, is whether the quest for a fixed and judicially unalterable legal standard for the protection of speech activity really is not, in the practical world, a hopeless and quixotic undertaking. The general issue is deeply complicated (we need not rehearse the arguments already given at the conclusion of chapter 3), but it is worth bearing in mind that the actual history of the First Amendment in this century does not provide much comfort for the view that judges armed with fairly clear mandates with which to insist on toleration of speech will in fact do so when the needs of intolerance flow within the society.[31] Whether "clearer," more certain standards would work better is, of course, the question, but it is important to bear in mind that virtually everyone has agreed that *some* exceptions must be created for social regulation of speech, which means that, at least in First Amendment jurisprudence, an "absolute" will always be a relative concept.

Quite apart from whether a "fixed," or inflexible, standard is feasible to formulate or would be effectively enforced, the question remains whether it would be worth it. We have considered a variety of harms a community may suffer when forced to tolerate speech activity within its midst, and such harm as does in fact arise in a given case must be balanced against the gains in added protection for speech that is potentially secured by the fixed rule. Hand's proposed incitement test, which has been widely interpreted as providing that speakers are invariably protected from regulation as long as they refrain from explicitly advocating violation of law, would in all probablity end up permitting the punishment of the feckless and ignorant (who wouldn't be aware of the subleties required to avoid the law) and prohibiting the punishment of those who might be truly dangerous in times of genuine emergency. Beyond that, we must also consider what alternative forms of intolerance will occur when intolerance in the speech area has successfully been blocked. The simple fact that legal coercion against speech activity has been checked does not necessarily mean it will not resurface in other forms. Because the free speech results are likely to be considered highly arbitrary and socially unacceptable, the intolerance may actually be stimulated—and become excessive—in other areas of social interaction.

The wish for certain tests and for fixed and permanent results in the free speech area is premised on a view that free speech is concerned almost exclusively with the limited aim of protecting what is thought to be speech of true value from legal censorship. In that view, protection of speech at all costs seems sensible, but not when the concern is with the general development of a capacity relevant to social interaction generally.

Thus, the more we take a wider perspective on the social functions of free speech, the more the matter of deciding on the relevant tests for it looks different. In fact, the view that free speech is related to the pursuit of a general capacity leads us to see litigation as an *opportunity* rather than a reason for distress.

Litigation provides the framework, the occasion, for the community to think about the things free speech is intended to raise for thought. If cases are dismissed with computerlike efficiency, such matters are removed from public discourse or, possibly, raised only in a context (the legislature) in which the considerations favoring censorship are perhaps more likely to be emphasized. Moreover, the litigation process provides the forum that makes it possible for those troubled by the posture of toleration (very possibly for good, or understandable, reasons) to express their rejection of the way of thinking manifested by the speech activity. We ought to encourage these people to articulate their concerns, so that their claims may be properly heard and considered and also so that the people themselves may be brought to a closer understanding of their own motivating concerns, for (to borrow once again a thought from the classical vision of free speech) by listening to themselves speak, they may better understand what they think. Time is essential to achieving that goal, and a flexible standard is essential to achieving time.

The ventilation function of litigation also makes achieving toleration and retaining the symbolic function of free speech more possible. Shifting responsibility for toleration in particular cases to the courts is important, as I suggested in chapter 4, but an important ingredient in that subtle process involving a partial transference of responsibility is assuring those who object to the speech an opportunity to clearly articulate their objections. Moreover, a kind of recapitulation function is importantly performed by the cases, as generation after generation must necessarily be instructed in the lessons offered by the free speech principle.

Finally, we ought not to underestimate the importance to any institution of having a good and clear idea of its purposes. One of the deepest, yet unnoted, difficulties with the long-standing debate over whether the First Amendment should be applied according to a balancing method or a "categorizing" method has been the lack of attention paid to the critical importance of the assumptions about the underlying purposes of the free speech principle that the participants bring to the debate.[32] The less clear we are as to those purposes, or the less persuasive they are to account for the free speech position being defended, the more difficulty we are going to have in applying the principle, whatever approach is adopted. When balancing takes the particular form of weighing the benefits of the specific "speech" in the individual case against the social "injury" sustained as a result of that speech on an ad hoc basis, it is properly criticized for being loaded against the courts' reaching a decision for toleration; and if one sees the "benefits" to be derived from toleration as those arising from this particular confrontation and as being, for example, the incremental gain to the store of public information or general knowledge, then one can readily see why such an inquiry is feared by the categorizers. But now the problem can be seen, properly, as turning on the conception we take of the institution of free speech. If we see the "benefits" of free speech as derived from its being a partial area of social activity in which a position of extraordinary tolerance is generally undertaken, the inquiry is not loaded against free speech at all. By shifting our idea of what we are trying to accomplish with free speech, we alter fundamentally the terms of the discussion about how it should be applied. Balancing looks dangerous depending on what is being balanced. By taking free speech according to its function of helping to create a tolerance ethic within the society, that method is both transformed and rendered more appealing.

In closing this discussion about a general approach to drawing lines in free speech cases, it would seem helpful to consider how we might approach a contemporary case of extremist speech under the standard just proposed. Again, let us draw on the *Skokie* litigation for these purposes. It is especially important to focus on a contemporary case like *Skokie* because the standard urged here demands that we maintain a close sensitivity to the conditions prevailing at the time of the controversy. That is not to say, it should quickly be added, either that all decisions are necessarily of time-bound significance only or that the principle may be easily abandoned when times get rough and the harms significant. We begin, under the tolerance principle of free speech, with an understood commitment to extraordinary self-restraint; coupled, therefore, with a willingness to be sensitive to context is the idea that the free speech principle requires us to begin with a strong presumption in favor of toleration, which can be overcome only after it is determined that the society has little or nothing to gain in the various ways that the tolerance principle proposes the society will benefit from self-restraint and, by comparison, a great deal to lose.

We have already considered (in chapter 4) several important ways in which the general tolerance function favors an insistence on self-restraint in the *Skokie* context. Quite clearly, this was an occasion on which the *potential* for excessive intolerance was present, which is all the courts should, or realistically can, be concerned with. Most people no doubt felt the fear of losing control, of acting with excessive intolerance, in a confrontation with the Nazis. Many must have felt the same troublesome and complicated feeling of identification hauntingly expressed by one Skokie resident: "I heard myself like Bull Conner opposing Freedom Marchers," he said.[33] *Skokie* was also a good case for highlighting these issues of tolerance and intolerance, for it attracted widespread public attention. In these ways *Skokie* was a good case for evoking the kind of personal and community issues, for confronting highly troublesome ways of thinking, that it is precisely the function of free speech to achieve.

But were the harms of toleration simply too great? It is difficult to give a full and clear answer to this question because, as mentioned in chapter 2, the courts in *Skokie* were unfortunately unreceptive to undertaking any full-scale examination of the harms of toleration. Not all the blame can be laid at the judicial doorstep, however, for the city itself did not press its case in a way that would best highlight the range of harms likely to be suffered as a result of tolerating the Nazi speech. There was testimony by psychologists on the possible physical injuries many Jews would suffer as a result of the evocation of painful memories by the speech. The likelihood of such injury, it was said, showed that this speech act was the equivalent of a physical assault, which demonstrated in turn, it was further argued, that the speech act was properly subject to regulation, as was any physical attack.[34] Whatever the validity of these estimates of the psychological and physical injury involved in the speech, the attempt to correlate that injury with the injury of nonspeech acts as a means of enhancing the claim for regulation only points up again the serious confusion underlying First Amendment analysis about the basis for tolerating speech activity. Free speech is not, or ought not to be, premised on an assumption that speech causes no (or even less) injury, even physical injury; we assume injury but further assume that greater societal benefits will be derived from the lessons learned through toleration.

Nevertheless, while we recognize that the record in *Skokie* is not so complete as we might like it to be on the issue of the costs of toleration, would our best estimate of the costs lead us to abandon the presumption for tolerance? Here I think we must recognize that, realistically, the costs did not involve the potential for immediate persuasion of grave and socially harmful acts. But were the harms sufficiently serious to be equated with those that lead us to exempt regulations of obscenity, fighting words, or libel? My judgment is that they were not.

In discussing the obscenity exception to the First Amendment, we considered the importance of arriving at a judgment on the matter of possible confusion about the basis of the choice to be tolerant. This seems not a substantial enough concern with the Nazi speech. While anti-Semitism is a problem in American society (just how serious would have been a potential issue in the case), it is not of such magnitude, or so pervasive, as to transform toleration into an act of implicit condonation. This society is not in this sense in a situation like that of Germany, where even today that society maintains extraordinary restrictions on Nazi symbolism, no doubt because of a fear of what would be implied by tolerance.[35]

Yet, can not it also be said of the proposed march in *Skokie* that it involved an attempt to inflict serious injury on only a few members of the larger society? Did not this make the case akin to libel and, therefore, properly exempted from the First Amendment? Again, I think the better judgment is against drawing a parallel. One of the things that made *Skokie* fascinating as a study in ambiguity was the degree to which the "messages" being communicated were so complex. We noted in chapter 1 that the explicit message the Nazis claimed they wanted to express was that they were being unfairly denied their free speech rights. We may put that aside, however, and focus our attention on what were quite clearly the messages everyone knew were intended to be conveyed. These messages were importantly mixed, too: They involved not just anti-Semitism but Nazism as a general political program, and the importance of that mixture of messages was that it made it possible for people to think of the speech as an attack on the entire society and not just the limited community of Skokie. In a sense, the larger society could absorb the interests of the smaller community into itself; it could identify with the harm suffered because it shared in the harm. In this way the injury was not isolated, the meaning of tolerance was made less confused, and important reassurances about the issue of anti-Semitism were made possible.

All these factors strongly mitigated the seriousness of the costs involved in the *Skokie* case. The opportunity for litigation provided a powerful forum for addressing the issues involved, and the judges were able to assume immediate responsibility for the choice of tolerance. On this latter issue it may perhaps have been true, though one could never say for sure, that many Jews half welcomed the command that there be tolerance—like the person who succumbs to the restraining hands of friends while attempting to retaliate against the speaker of a personal insult.

## II

Let us now look at another group of basic First Amendment issues, and at one especially prominent perspective on those issues, as a means of illustrating how the understanding achieved under the tolerance principle can assist us in developing doctrine. We will be considering regulations regarding the time, place, and manner of speech activity and regulations regarding expressive nonverbal behavior, customarily referred to as symbolic speech. If the society seeks not to prohibit some idea or expression, but only to regulate the circumstances under which that idea may be expressed—the time, place, or manner of the expression—how should such a regulation be regarded under the First Amendment? If, additionally, the society seeks to restrict nonverbal behavior, like wearing an armband or burning a draft card in protest against a war, as occurred during the Vietnam era, should the First Amendment have anything to say about that kind of regulation? What we shall find is that the general tolerance theory of free speech provides us with an organized way of thinking about these problems, as well as with the foundations for reaching a judgment in particular cases.

Consider first the time and place regulation. Once again, it seems fairly clear, and no one has really ever contested this, that some limit on the timing and location of speech activity has to be permitted. None of us could bear an unrelenting barrage of speech we dislike; some escape must be recognized and some protection offered for the numerous activities we legitimately wish to pursue but which we could not if others could interrupt us with their harangues. In substantial part, the system of public and private property provides a fairly built-in and workable compromise, though it must be supplemented to some degree with additional regulations (such as restrictions on loudspeakers in neighborhoods). Thus, the system that has evolved makes good sense. Essentially the cases provide that in the public areas of cities—on public sidewalks, in parks and the like—speech activity must be left relatively undisturbed; but the home is another matter and greater insulation from speech is allowed there.[36] People must be able to escape political dissent, and it seems to be the best solution for everyone concerned if we designate a segment of the community as being relatively open for speech activity while leaving others somewhat more closed; a more evenly modulated system would probably yield less for everyone—and here we see, within free speech doctrine itself, a reflection of the beneficial uses of a mixed system of regulated and unregulated areas.

Under the general tolerance perspective, the key concern in these sorts of cases is that we give those wishing to confront us with unpopular speech activity a serious and meaningful opportunity to do so. What constitutes a serious and meaningful opportunity is, of course, a highly debatable point. Although we recognize that many difficult judgments remain to be made, it is no small achievement to have defined the general framework in which we will ultimately have to make such judgments. One of the most serious problems with the doctrine of time, place, and manner has been the absence of any clear sense of how courts are to analyze the cases under it, beyond assessing whether the "free speech interests" outweigh the "social interests" being furthered by the regulation (which typically involve things like preserving peace and quiet, litter-free areas, and uninterrupted traffic flow). The general tolerance function of free speech helps supply the definition of the "free speech interests" at stake and, therefore, gives a perspective from which a balance can be more rationally reached.

This leads us to another defect in the traditional analysis of time, place, and manner in free speech cases. This is the sense conveyed, usually implicitly but sometimes quite explicitly, that such regulations are, generally speaking, far less serious intrusions into the interests served by the First Amendment than are regulations that seek to restrict speech activity because of its "content," as it is commonly referred to. To a degree this is true, but it is seriously overstated and the adverse consequences of thinking that way are not insubstantial. If free speech functions as a point in social interaction in which we seek to learn something about the intellectual character we will bring to most, or nearly all, social interaction, the central concern ought to be with the circumstances under which those lessons are likely to be learned; that may or may not lead us to prohibit all content regulation while being far more lenient toward regulations that limit the occasions on which the confrontations are likely to arise. The position we commonly take is not, by any means, self-evident. A frequent and full confrontation with some offending speech may be more beneficial than is more limited contact with all speech. We do a serious disservice to the First Amendment, then, by beginning with the assumption that time, place, and manner regulations are automatically a second-tier problem, one less troublesome from a First Amendment perspective.

I have yet to say anything about "manner" regulations. The reason is that the concept is simply too encompassing, too broad, to be handled in the same breath with time and place regulations (both of which can be subsumed under the concept of manner). The primary problem is that a manner regulation may impinge on more than the circumstances under which speech activity can occur; it may cover, or apply to, the speech activity itself. For example, in *Cohen v. California* the state argued that Cohen's inscription of Fuck the Draft could be prohibited because the first word was deeply offensive to many and because only the manner of speaking was being regulated. Cohen, it was said, could easily find another way to express his view of the draft. To this argument, of course, Justice Harlan responded negatively, saying that it was impossible to separate the "way" in which words were put together and the message actually conveyed.[37] This particular term, he added, was important to the expression of Cohen's depth of emotional feeling about the draft, which was also a legitimate and an important element protected by free speech.

Harlan's conclusion was right, but the reason he gave was not entirely satisfactory. One could argue that he was incorrect on the basic point, contending that it is, in fact, possible for someone to express *the same* idea by using different language. That is a difficult challenge to meet, but it is not the point I wish to make here. Rather, I would say the problem lies in associating the "message" of the four-letter word with the message about the draft. It seems just as likely (actually, I think, more probable) that Cohen was making *many* different statements, some of which would have nothing whatever to do with the draft. To find the linguistic utility of the term *fuck* by its association with the issue of the draft is to interpret the possible statements emanating from Cohen's jacket (and accompanying behavior) much too narrowly. He may very well have been using the word in such a way that it would challenge and offend people who held a variety of attitudes and values that he disagreed with. The use of the term, therefore, was a deliberate act of provocation, like an obscene gesture or a push.

The reason for making this point is to emphasize the need to recognize the complexity of the audience and speaker interaction in any speech activity. Intolerance usually does not arise because of the expression of an abstract idea, but because of the way in which it is expressed, which in turn reveals a way of thinking communicated through the speech act. The term Cohen used is not objectionable to millions of people; in fact, many no doubt have used it themselves on occasion. What does matter is the context in which it is used, which will reveal many things about the speaker—how the speaker is thinking and what the speaker is "saying" to his or her audience. As in the determination of what is obscene, the answer is typically to be found in the mind revealed through the speech act. Reading a fairy tale, Willard Gaylin reminds us, can be done in such a way that we feel it is obscene.[38] The same, however, is also true of intolerance toward nonspeech behavior, which brings us to the area of symbolic speech.

A decade ago Dean John Hart Ely analyzed the time, place, and manner doctrine and symbolic speech cases in a major, influential law review article.[39] In that article he contributed several very important insights for First Amendment analysis. He used *United States v. O'Brien*[40] as the pivotal case for his analysis, a case involving the prosecution of O'Brien for failing to have possession of his draft card, which he had burned in protest against the war in Vietnam. Ely developed the thesis that what really mattered in every First Amendment case was whether the concern behind the law being challenged as unconstitutional was over what he called the "communicative impact" of the speech. By communicative impact Ely seemed to have in mind the traditional notions about what speech could do, that is, persuade others to commit acts or to offend the sensibilities of hearers. In any event, to Ely the "critical question [was] whether the harm that the state is seeking to avert is one that grows out of the fact that the defendant is communicating, and more particularly out of the way people can be expected to react to his message, or rather would arise even if the defendant's conduct had no communicative significance whatever."[41]

If the government's motivation in restricting the speech was with the "communicative significance" of the act, then, according to Ely, the law would have to be held unconstitutional unless it was determined to fall within one of the limited categories of exceptions to the First Amendment (obscenity, libel, and so on). If, on the other hand, the law was concerned with a "noncommunicative" consequence, such as the maintenance of a convenient system of draft records, then only a "balancing" standard was to be employed.

Now, one important consequence of this method of analysis was to collapse the time, place, and manner doctrine and the question of what to do with nonverbal acts of expression into a single inquiry. The point is no longer to ask whether, for example, this is a regulation of the "timing" of speech, but rather whether the regulation of the timing is directed at forestalling the communicative impact of the speech. Similarly, the question is not whether this is a regulation of "expression" or "action" or of "speech" or "conduct" (both of which courts and scholars had previously proposed as methods of inquiry), but rather whether the regulation is directed at the communicative impact of the communication inherent to the conduct or of something else.[42]

Ely applied this analysis to a number of cases: *O'Brien* had been properly decided; at least the Court there had applied the proper standard to the regulation at issue. Since the purported legislative motive behind the regulation had been directed at maintaining an efficient system of draft records, and not with a concern over the impact of the antiwar message being communicated, it was a second-level First Amendment issue: "The interests upon which the government relied were interests, having mainly to do with the preservation of selective service records, that would have been equally threatened had O'Brien's destruction of his draft card totally lacked communicative significance—had he, for example, used it to start a campfire for a solitary cookout or dropped it in his garbage disposal for a lark."[43] On the other hand, *Tinker v. Des Moines*, the case involving a high school regulation forbidding the wearing of black armbands because of the "disruption" such clothing was likely to cause among students, was a first-level case because the concern behind the regulation was to forestall the reaction of those students who objected to the message being communicated by the wearing of the armband. The same was true for *Cohen*: "[T]he critical point in *Cohen*, as in *Tinker*, is that the dangers on which the state relied were dangers that flowed entirely from the communicative content of Cohen's behavior. Had his audience been unable to read English, there would have been no occasion for the regulation."[44] As for something like a regulation prohibiting defacement of public buildings, this should properly be treated as a balancing case because "there a governmental interest quite obviously unrelated to the suppression of expression is implicated, namely, the cost and trouble of sandblasting."[45]

The problems with this method of analysis begin with those I have already mentioned in the discussion of the time, place, and manner doctrine. In particular, there is a need for some theory to explain what we are to balance when balancing is supposedly called for. (There is also a need for a theoretical explanation as to why regulations falling in the first level should be treated in the way proposed, a method of analysis I have already suggested is excessively rigid, as well as to why we have the exceptions we do for regulations of this kind.) Additionally, we encounter here again that implicit indication that regulations on the second track (not concerned with communicative impact) are somehow significantly less troublesome than those on the first. This, however, leads us to a more subtle and complicated difficulty that must be clarified.

Consider Ely's suggestion that regulating behavior like writing graffiti on public buildings is appropriately treated as something less serious, from a First Amendment standpoint, because the concern behind the regulation involves a desire to protect a noncommunicative impact interest, namely, the avoidance of the expense of sandblasting. The primary problem with this view is that it minimizes the complexity of the motivations behind the adoption and enforcement of the regulation, making it seem more innocent and unproblematic than it potentially is; more specifically, it ignores the fact that such a regulation is potentially infected with the very same elements of excessive intolerance that are encountered in ordinary confrontations with troublesome speech activity. The net effect of overlooking this connection is to stop the transference in meaning between the free speech principle and other social regulation and interaction, which, of course, is absolutely vital to the free speech enterprise.

A major part of the difficulty here stems from our failure to see that the "expression" involved in the act of writing graffiti is of several different kinds. When someone inscribes Get Out of Guatemala on the front of a public building, that person is "saying" many things besides the recommendation that the society withdraw from Guatemala—just as we saw was true with Cohen and his jacket. The graffiti writer is "expressing" disrespect for authority or for public property and may be "saying" something about the writer's attitudes toward those who would prefer a neater aesthetic environment, about "bourgeois" norms and the like. There is often an underlying attitude of hostility or contempt in such an act. It is a destructive act, which is why a policy of leaving some chalk and a blackboard in lavatories can never be an effective solution to the graffiti problem.

Even more important, we must understand that these underlying feelings—or ways of thinking—which are reflected and communicated in the act of defacement, will unquestionably affect our (or our representatives') responses to the offenders; that is to say, our responses will potentially be troubled by the very same conflicts and difficulties we encounter in trying to arrive at a proper response to any simple speech act (for example, the use of indecent words). In fact, it would be better, and more accurate, if we were to describe the basis of the motivation behind intolerance of speech activity not as concerned with communicative impact but with the way of thinking manifested by the act. What stirs people to want to punish a speaker is often not the explicit message at all—we may agree with the speaker that the draft is a miserable social policy and ought to be abolished—but what we call (to give a few examples) the "insensitivity" or the "thoughtlessness" or the excessive "hostility" manifested in the way in which the view is expressed; and precisely the same is true with nonverbal acts—the way of thinking behind them colors our response to them.

To suppose, therefore, that in dealing with graffiti writers, our motivations can be reduced to a concern over having to incur the expense of sandblasting is to ignore this inevitable interweaving of concerns about the mind and general attitude of the offender. The point is not that avoiding unnecessary expenditures is not a concern, or that the presence of such a factor (that is, one not concerned with the mind behind the act) may not be an appropriate point at which to draw the limits of free speech, but rather that it is vital we bear in mind that the response we give when the factor is present will virtually always be infected by the very same feelings that so trouble us when we deal with "speech" exclusively. There is no significant time in social regulation when there is any sharp line between concern over the communicative impact of behavior and any other concerns, and it is precisely because of that reality that the principle of free speech has such enormous potential meaning for us.

The central problem, then, with the hypothetical—and what makes it dangerous for our thinking about free speech—is that it conveys the impression that if we are not concerned with the message Get Out of Guatemala, we are not really concerned with the communicative impact of the behavior. What is initially required, therefore, is a fundamental restatement of the test: we should balance, not when there is no concern over the way of thinking manifest in the act (or, in Ely's terms, the communicative impact of the messages), for that is rarely if ever the case, but when there are concerns in addition to that over the way of thinking expressed (which will usually involve direct physical injury to people or property).

To say this is to bring ourselves to the point of a dilemma in our attempt to define the parameters of free speech. The dilemma arises in this way: We began our analysis with a kind of tacit assumption that the feelings motivating intolerance of speech activity are both unique and fundamentally bad, essentially unworthy of our consideration or respect. As we come to see that the same considerations are often at work behind our reactions to nonspeech behavior, we are able not only to see the wider symbolism of what we do under the free speech idea but also to see the potential reasonableness of those feelings even toward speech acts. The connection both stimulates potential meaning, as we found through the discussion in chapter 4, and highlights the self-imposed sacrifice of legitimate interests as we pursue that meaning.

In a sense the decision about what type of behavior will be selected for the special, extraordinary toleration we commit ourselves to under the free speech principle is fundamentally involved with pedagogic concerns. No line can be drawn on the basis of a judgment about the *moral* legitimacy of the motivations behind our intolerant responses to different kinds of behavior, nor can one be drawn on the basis of a difference in the *degree* of harm generally sustained when speech and nonspeech acts are tolerated.

Still, under the free speech principle as we think of it today, there does seem to be a working consensus that we should treat regulations exclusively concerned with avoiding the harm people suffer when they confront the speech acts of others as the primary category of behavior covered by the free speech principle. So long as we recognize that genuine harm is being suffered as a result of that choice, and that the same harm is suffered with nonspeech behavior too, we may reasonably proceed with an analysis that consciously accords less weight to the *same* type of harm when that harm is inflicted through a speech act than when it is inflicted through a nonspeech act *and* the society has moved against the nonspeech act to avoid additional kinds of injury.

Still further reflection, however, suggests yet another possible basis for separating speech acts from nonspeech acts for these purposes—one that focuses on the more powerful *communicative* potential of nonspeech behavior. Words are often regarded as a cheap means of purchasing an identity and, therefore, as not providing a true "demonstration" of the speaker's beliefs or values or an opportunity to reject those of others. Here we encounter another example of how in the free speech area every argument can be turned against itself. Just as we saw in chapter 2 that being free to impose *legal* restraints on speech acts is a more effective means of demonstrating community values than is any mere verbal declaration, and likewise in chapter 4 that actually engaging in extraordinary self-restraint toward speech acts is a more effective demonstration of the tolerance function of free speech than is any simple verbal declaration to that effect, so too we might expect that the needs of belief may incline many speakers to turn from words to more serious forms of actions thought to be more suitable to show the very attributes Holmes in *Abrams* saw in the "logic" behind intolerance. In fact, we might think of free speech as a means of channeling this need into a generally less harmful form of behavior (compared with the kinds of behavior people are likely to turn to as alternatives). Still, there will always be a natural desire to turn to nonspeech behavior, even (perhaps especially) those that inflict injury, as a means of more forcefully expressing what the "speakers" would like to say.

What is interesting about this move is that the expressive implications also tend to escalate, which is, of course, why unstable societies feel the need to react swiftly and forcefully to suppress any act that is perceived as a step in a projected revolution. In this way, reserving the special free speech presumption against regulation primarily for those concerned with avoiding the harm of speech acts may be thought justified because of the more powerful communicative impact of nonspeech acts (to return for the moment to Ely's terminology).

This insight has implications not just for the problem of how to deal with regulations concerned with both nonexpressive and expressive harm (as is true in graffiti cases) but also for the problem of what to do with regulations of nonspeech behavior that are concerned exclusively with expressive injury—as, for example, would have been true in *O'Brien* if the regulation had been motivated by a desire to curtail antiwar expression. The point that must be considered is whether free speech ought to follow a more lenient approach to these types of regulations because of the likelihood that such acts will produce greater social injury. With behavior like burning draft cards (or, to take another example, wearing black armbands), it seems unlikely that significantly greater injury will occur, though we ought to consider not just the implications of the particular act but also the possibility that other acts may follow a move from speech activity to nonspeech activity as a means of dissent (burning military installations, to take an example that may have been relevant to the Vietnam era). Also, our views of such regulations are affected by how specifically the regulation is tailored to particular viewpoints—the difference between a ban on all "political" expression through defacement and a ban on antiwar expression.

As a general matter, it seems unlikely either that regulations of this kind are likely to arise or, more accurately, that courts will ever really be able to ascertain whether the motivation behind the regulation is exclusively concerned with expressive harm, since legislatures are not likely to admit openly to such a motivation and will use other concerns as camouflage (as probably occurred in *O'Brien*). Uncovering "real" legislative motives is no simple task. In any case, the structural implications for the respective roles of the judicial and the legislative branches have always led a large segment of the legal community to resist movements in that direction.[46] What should happen is probably this: restrictions on nonspeech behavior that are rather obviously concerned exclusively with expressive injury are treated as within the regular domain of the free speech principle, while others that are less clear are let go. Still, even from a strong First Amendment standpoint there is something to be said for drawing the boundary of free speech at the point of the speech act.

How should we decide how much nonspeech activity will be required as a constitutional matter in order for speech activity to have a chance to play the role it will under the general tolerance function of free speech? As indicated a moment ago, we use the system of property to help us achieve a rough balance.

But we add to it the idea that *public* property must be accommodated for some speech activity. Again, as observed earlier, the standard we should be applying in determining the scope of that accommodation is whether a genuine opportunity exists for confrontation. Obviously that is only a starting point, but it is an important one, partly because it indicates the direction of other considerations as well.

Streets, public parks, sidewalks, have been opened to speech activity.[47] What about the sides of public buildings, or their insides? These, as Ely's graffiti example indicates, have not been turned over to general speech activity. Is it because we value the cleanliness of buildings more than the peace and quiet of our public parks? I think not, except in an indirect sense. Here the solution seems to me to rest in the public nature of the speech activity in question. The problem with graffiti is that they are done secretively, like an obscene telephone call. Like that sort of behavior, the messages that tend to be communicated are frequently of the most offensive and troublesome variety. What is most worrisome about secretive behavior is the greater likelihood that people will indulge their impulses to act in this manner, which, when tied to the greater expressive injury of the destructive act, makes a strong case for allowing regulation. The anonymity of the act makes the vast web of social, unofficial constraints and penalties ineffective, and as a consequence there will be too much of the behavior and too much social injury. With free speech we wisely use natural curbs whenever we can, and limiting free speech (perhaps not entirely but nevertheless largely) to *public* speech acts provides a natural and desirable degree of containment.

# 7
# Searching for the Right Voice

It is important that we consider how we think and talk about free speech in the context of controversies over the limits of the principle. A general study of free speech rhetoric would be a rewarding undertaking in itself. We would want to consider, for example, why the area has attracted what seems like a disproportionate share of the most beautiful writing to be found anywhere in the law. In fact, it would not be out of the question to regard free speech as fundamentally a literary enterprise. The law's foremost rhetoricians—Holmes, Brandeis, Chafee, Kalven, and others—all brought their remarkable verbal skills to bear on the free speech problems of their time. The dissents of Holmes and Brandeis in those early cases, in which the groundwork was laid for the development of the modern free speech idea, possess a rhetorical beauty that draws us back again and again to their rhythms. Kalven himself remarked on the "almost uncanny power of these dissents";[1] later, Bork said of Holmes and Brandeis—whom he described as "rhetoricians of extraordinary potency"—that "their rhetoric retains the power, almost half a century later, to swamp analysis, to persuade, almost to command assent."[2] But while Bork regarded this verbal charm as something of a siren's song toward which we had better be wary, even turn a deaf ear, we need not, and ought not, be so dismissive toward the phenomenon of beautiful writing as we set out to understand the contemporary significance of the free speech principle. In fact, this characteristic of the area seems yet one more piece of evidence for the observation that we ignore at the peril of our own ignorance the tremendous symbolic role played by the First Amendment in this society, for cultural symbols typically attract writing of this caliber and power.

There exists an especially powerful—indeed, one should say, imperative—reason why we should undertake a critical self-examination of the thinking and talking that go into defending free speech against attempts to limit its application. At this point in the development of the thesis about the relationship between free speech and social tolerance, that reason should be fairly apparent: If the whole point of the free speech enterprise is to force us to become conscious of the need to control various impulses we feel when confronted with other people's thinking that we find objectionable, for one reason or another, we must recognize that free speech thinking itself may fall victim to those very same impulses.

Anytime we undertake the task of settling a dispute, the way we go about performing that task—in particular, the way we treat the parties and their respective claims—will set an example for those persons, and others who may be observing, of how they ought to treat each other and how they ought to think about issues generally. How a case is decided can be just as important to the society as what is decided. In free speech controversies there is an added twist to this role of process, one that profoundly alters the balance in favor of the importance of methodology, because the peculiar feature of the free speech principle is that the very incapacity it seeks to overcome can infect the actual application of the principle itself. In this sense, a free speech case will always be directly exemplary of the idea it stands for and seeks to inculcate. It will exist as a kind of miniature, a controversy just like any other in the world, which, like the others, is subject to the same distortions of intellectual biases by the participants in the controversy.

The simple truth about free speech is, therefore, that its proponents face the risk of exhibiting precisely the intolerant mind that the principle is intended to point up and condemn. An intolerant defense of tolerance is more than just an anomaly; it may cancel out, even reverse, the gains the society hopes to achieve through the institution of free speech. We must teach ourselves, therefore, that the pursuit of tolerance is not necessarily an inoculation against the malady of an intolerant mind.

This is the primary theme explored in this chapter. We shall also find that through free speech the focus on the nature of the intolerance impulse provides us with a basis for thinking about a variety of issues connected with the application of the principle itself.

## I

In chapter 3, which describes the fortress model, we noted how fears about the oppressive tendencies of democratic majorities tend to foster a certain defensive style of discussing and defending free speech. This occurs as part of a strategy for presentation of the principle. An underlying aim of this way of thinking is to construct a mental world in which it is unthinkable to consider making exceptions to the free speech principle. To this end a variety of rhetorical techniques are employed: The history and text of the First Amendment are treated as fixed and unbending; prior cases are regarded as unyielding of exceptions and lines as impossible to draw.

The techniques employed, however, transcend the construction of seemingly unalterable legal authority. They involve an array of methods used to frame the terms on which disputes of this kind will be thought about. These methods are often subtle and hard to identify at first, but once they are identified, we see how powerfully they function to constrain appreciation of the complexity of the issues faced.

In the characterization of tolerance and intolerance as opposing ends of a spectrum of good and evil, the former is associated with fearlessness and courage, the latter with timidity and weakness. Such a way of talking about intolerance also blends into a series of implicit assumptions about the limited harmfulness of speech for those who must tolerate it under the free speech principle. In chapter 1 we noted Wigmore's objection that Holmes had minimized the need of the country to prohibit speech like that in the *Abrams* case, and in chapter 2 we elaborated on that objection. Speech harm, it was argued, is narrowly viewed as involving only a risk of immediate persuasion or an element of offensiveness for the audience—both of which are further minimized by assuming that any "action" the speech induces may be punished and that the offense taken may easily be avoided by "averting the eyes." Furthermore, the social harm actually considered is only that arising from the speech act in the particular case, and the burden of explaining that harm is placed on those who wish to limit the speech.

On the other side, still other conceptions are used to skew the balance of considerations in favor of toleration of speech. Here we encounter the language of "rights," which, as I have argued previously, seems often intended to operate as a form of closure on further discussion, as well as on having to provide reasons in favor of toleration. As to the speech act itself, we frequently encounter a tendency to minimize it as potentially bad behavior. Speech is not viewed as the possible instrument of an intolerant mind. When recognized as bad behavior, the speech act is usually excused as natural or inevitable. It is part of the standard free speech rhetoric to quote Madison's statement: "Some degree of abuse is inseparable from the proper use of everything; and in no instance is this more true than in that of the press."[3] In *New York Times Co. v. Sullivan*, the Court quoted with approval an earlier case, in which the Court had observed that "sharp differences arise" in religious discussions and that "to persuade others to his own point of view, the pleader, as we know, at times, resorts to exaggeration, to vilification of men who have been, or are, prominent in church or state, and even to false statement," and then added approvingly Mill's similar perspective on bad speech behavior:

> [T]o argue sophistically, to suppress facts or arguments, to misstate the elements of the case, or misrepresent the opposite opinion... all this, even to the most aggravated degree, is so continually done in perfect good faith, by persons who are not considered, and in many other respects may not deserve to be considered, ignorant or incompetent, that it is rarely possible, on adequate grounds, conscientiously to stamp the misrepresentation as morally culpable; and still less could law presume to interfere with this kind of controversial misconduct.[4]

Speech, therefore, is viewed as something that can be "abused" but, importantly, not as something that can spring from the same motivations and thinking that lie behind the act of suppression of speech, which, of course, we are quick to condemn—or, for that matter, behind the way in which we deal with the act of suppression of speech.

How we react to speech activity is a most complicated issue, as the discussion up to now indicates. But it would be a most unfortunate result if people came to believe that all intolerance of speech was a bad thing. We have noted before that our willingness to tolerate troublesome behavior often depends on the context in which the behavior occurs. We may, for example, properly be willing to expect less of dissenters than we do of those who feel themselves part of the ruling group. But even with dissent we believe, or ought to believe, there are boundaries of bad behavior that should not be crossed. At some point we will insist that people reform themselves, however "natural" it may be for them to want to behave in that way. As we have seen elsewhere, it is always a tentative, even wavering, line between when we will accept people as they are and when we will expect them to curb their inclinations to misbehave. What is critical is that we understand not simply that in "the realm of religious faith, and in that of political belief, sharp differences arise," but that the faith and beliefs that lead people to resort "to exaggeration, to vilification of men . . . and even to false statement" also lead people to engage in the censorship of speech, which we deplore, as well as a host of other acts of excessive intolerance—which, again, may include how those censoring speech are treated under the free speech principle itself. Fanaticism is a state of mind with unlimited possibilities for affecting behavior.

The upshot is simply this: We can criticize, as we have criticized, all the methods just recounted for dealing with free speech controversies as illogical, manipulative, or even ineffective (as Chafee, for example, said of the rhetoric of "rights"[5]), but it is an additional and important step to recognize them as manifestations of the very same impulses that so trouble us when we encounter them in the minds of those who censor. It is like lying while enforcing a principle of good faith.

To some degree the failures in the traditional rhetoric about the First Amendment can be chalked up to limitations imposed by the language we have inherited for talking about the free speech principle. A distinct eighteenth- and nineteenth-century tone characterizes much modern discourse about free speech. The official language of the First Amendment derives, as we saw earlier, from the Enlightenment era, and so we frequently encounter heavy doses of talk about the tyrannical tendencies of governments and the rationality of people. The point is not that contemporary reality has drained these notions of all meaning, but rather that the issues facing the present world are just that much different from those facing the nation when it was still on the brink of shifting to a democratic system of government. The society has moved beyond the stage of trying to secure the basic forms of democratic sovereignty, such as the power to discuss public issues and to elect representatives to vote on those issues. But the use of this inherited language tends to cloud our ability to see newer social functions emerging from the free speech idea.

Are we, perhaps, better off if the newer functions are left to exist just beneath the surface? Perhaps we are better off treating free speech as something of a social taboo, as essentially a one-sided issue, with nothing to be said against it and little to be said for those who would chip away at its exterior.[6] Taboos necessarily involve distorted, one-dimensional thinking, but they have the distinct merit of making certain undesirable behavior less likely to occur. In chapter 4 I suggested that an advantage of the system of judicial enforcement of the free speech principle is the ambiguity about who is "responsible" for its application in specific cases, which might suggest that the principle would benefit even further from having an atmosphere in which exceptions were unthinkable. Therefore, it may reasonably be thought, if we make free speech into a social taboo, a sacred symbol, we can accomplish the preservation of both the symbol and its symbolic meaning for other areas of social interaction.

But it is precisely this assumption about the continued possibility of valuable symbolic meaning for free speech that seems most doubtful if the principle is to be treated as a taboo. If everyone understood, at some level, what was going on, perhaps it might work. But that seems unlikely. What is more likely is that the typical techniques of a defensive posture, of trivializing the opponents' positions and smearing them with bad associations, are likely to antagonize those who feel (rightly) that regulation is not unreasonable; and just as confrontations with intolerant-minded speakers may stimulate an intolerant response to the speakers' messages, so it may happen with the response to the free speech principle. The cycle of intolerance will not be broken until or unless those who defend free speech break it themselves.

It seems doubtful, therefore, that the real purpose of free speech, of creating a context in which impulses to excessive intolerance are highlighted in a meaningful way, will be appreciated if the principle itself fails to abide by its own injunctions. The taboo mentality seems to work against, even to deny, the principle's larger meaning. We see this in the free speech characterization of speech activity as something capable of being abused but not as manifesting the intolerant closed-mindedness that we try through free speech to overcome. We can see it in the tendency to view free speech as concerned exclusively with a set of peculiar harms arising from the regulation of speech, while the vast area of the regulation of nonspeech conduct is treated as something fundamentally different or as justifying why we afford special protections to speech. The problem is essentially that we are unlikely to see both the true basis on which regulation of speech is restricted and the connection between what is potentially bad about our responses to speech acts and our responses to other acts, because to point up that connection would be to discredit our own behavior in implementing the principle. There exists, therefore, a powerful disincentive against making any connection between free speech and the larger arena of social behavior.

In earlier chapters it was argued that a precondition to understanding the new social meaning of free speech was to recognize the universality of the intolerance impulse manifest in the censorship of speech activity; in a sense, that is the point being developed here. We must also, however, be empathetic to the needs of intolerance as both inevitable and, actually, desirable. Perhaps a good way to do that is to acknowledge the convergence of the thinking that may motivate the censorship of speech and the thinking that may underlie the defense of free speech. The thinking in both instances may be constructed out of the same building blocks of premises about human nature and about law and may be also exposed to the same potential bias or deficiency in that thought process.

In both areas the concern is usually with the way of thinking behind the act and, significantly, with the potential of that way of thinking for affecting future behavior in a variety of ways. The act puts our own identity at issue, and law becomes an important source of symbolic expression about that identity.

As Holmes said, when you want to believe something very much and you fear losing it, it is perfectly natural to respond by suppressing the perceived threat. Law, moreover, gives us the opportunity to avoid particularized treatment of cases. Words cannot always be trusted to reflect what we believe to be the truth, or the depth of our feelings. Reason may fail us at the particular moment when it is needed most, either because our skills will, for whatever reason, not be up to the task or because others, perhaps momentarily swayed by passions, will have their judgments deflected from the truth. Sometimes, moreover, even to consider an idea—especially one we think is dangerously wrong—can lend to that idea an element of credibility, making it thereby more thinkable than it might otherwise properly be, and making ourselves the unwitting instruments of our own loss. Furthermore, we may naturally be inclined to think, both hyperbole and understatement are useful methods of arriving at the actual, more modest goal we seek. Mischaracterization is inherent to negotiated conflicts. Finally, we may fear that a too-balanced and tempered view of things will deprive us of the requisite will to spring to the defense of the principle at those moments when it is most in jeopardy, during periods of intense social stress when people succumb so easily and unwisely to their destructive impulses, just as it was suggested in chapter 6 that free speech must be prepared to accommodate the needs of belief in times of national emergencies, like a full-scale war.

While these ways of thinking are not, as such, entirely unreasonable, they may get out of hand and become excessive, just as they can in the contests over the regulation of speech behavior. The temptation is in that direction partly because it is the easiest route to follow and partly because we are uncomfortable with ambiguity and the possibility of change. What we can and should strive for is both a sympathetic understanding of the needs of intolerance, as well as a wariness toward them, and an awareness of their role in the enforcement of the free speech principle itself.

It would be desirable, therefore, if those who defend and apply free speech—especially, of course, as litigants and judges—viewed it as a central lesson of free speech that they themselves be wary of their own tendency to oversimplify, or, in effect, to censor, the complexity of the problems involved in the cases they deal with. The idea of free speech should reflect a constant appeal to reasonableness. The temptations against this are many and strong. The adversary context of the litigation system greatly accentuates the problem, for the tendency is, when we are under attack, to dig in and become even more of a believer in the position we started to defend. This is one of the most serious occupational hazards for the lawyer, for it is only too natural to sacrifice intellectual integrity by overidentifying with the client's position, by coming to believe as absolute truth the arguments made on the client's behalf. The pain of balancing the role of the advocate with being a separate person swamps the intellect. So, too, it is for the judge, who is also inclined to respond to attacks from litigants and dissenters by entrenching belief still further and by pronouncing, even if only implicitly, a doctrine of judicial infallibility.

Actually, the intellectual history of the free speech idea in this century provides a striking illustration of this reluctance or inability to remain open-minded and self-critical when others are expressing doubts and objections to the position being taken. Not until the last decade has any significant attention been paid in First Amendment scholarship to examining critically the premises underlying the free speech principle—whether free speech makes sense in light of what is done with respect to nonspeech behavior and whether vesting the interpretative and enforcement power over free speech in the judicial branch can be both justified and fairly judged a success.[7] Earlier scholarship, when reread from the point of view of the contemporary and seemingly so much more sophisticated analyses, often seems superficial and surprisingly unscholarly. At least to a degree, the difference may be seen to rest in the social differences in the periods in which the writings arose. We live in a time (as Bagehot said of late-nineteenth-century England) in which tolerance seems to come easily, in which the reactions to speech activity are not so troublesome as they have been at various times in the past, as in the 1920s and again in the 1950s. These were times in which free speech proponents felt that the principle was under fierce attack, and the chosen response to this reality was that of the advocate, of a believer.

While surely not an easy task, it would nevertheless be best if those who work in the free speech area tried to remain conscious of the tendency to revert to a more or less closed-minded posture in free speech disputes. One obvious means of avoiding the hazard we are considering is to follow the rule—which, admittedly, is as difficult to apply as it is simple to state—of giving genuine consideration to arguments for allowing suppression, not only in the process of deciding whether the principle should be applied in this particular dispute but also in determining how to present the decision reached.

Consider once again the decision of *New York Times Co. v. Sullivan*. Citizens must feel free, a majority of the Court held, to speak their minds about public officials, without fear of reprisal, whether through fines or libel judgments, because in a democracy public discussion performs a vital function of permitting citizens to exchange information and thereby arrive at better social decisions. Putting aside the variety of objections that might be raised to this as a matter of First Amendment theory (which we considered earlier), and taking the decision on its own terms, we can see serious objections to this way of presenting the Court's new decision about libel law. Speaking of what an opponent to *Sullivan* would argue, Dean Harry Wellington describes the problems well:

> The candidate, to the contrary, begs the court to recognize that the lies published about him misled the voters and thereby injured

> the political process. The statutory standard of due care, he insists, is the ideal standard for ensuring that the public is informed, rather than misled. Negligence is not to be encouraged in the reporting of political news any more than elswhere, and if due care costs more than carelessness, the purpose of the First Amendment requires that newspapers rather than voters should bear that cost. Moreover, if newspapers are free to lie, some of our most capable citizens will be deterred from running for office; the risk to reputation may outweigh the charm of public life.[8]

These are the kinds of difficult considerations that make the *Sullivan* result less than self-evidently correct. Sweeping generalities about the "profound national commitment to the principle that debate on public issues should be uninhibited, robust, and wide-open," while in some sense appropriate, must not be permitted to obscure the reality that the analytical and empirical questions underlying the position taken are highly complicated. The recognition of complexity ought to be the first rule, therefore, of effective free speech application. Judges may distinguish themselves from other decision makers in the degree to which they are able to engage in that recognition.

In an article assessing Black's contribution, as a justice, to the First Amendment, Kalven said of him that he possessed a fundamental prerequisite of a great free speech jurist: he had "passion" for the principle.[9] Although one understands Kalven's thought and applauds the courage needed to apply an unpopular principle, still, the statement ought to make us uneasy. Yet the troublesomeness of the compliment Kalven paid Black is softened considerably by the fact that no free speech analyst has been more representative of the tolerant mind than has Kalven himself, a quality that shines through his writings and probably is even more important, in the end, than any of his explicit messages.[10]

## II

There is a second source of pressure that distorts thinking about free speech, and it too becomes more vividly apparent to us once we have observed the relationship between the free speech principle and the search for a general capacity for tolerance. Earlier, in tracing the variety of social contexts in which exercising self-restraint toward one's own beliefs is required, we noted how important this capacity was to the identity and character of the judge. Probably nowhere else in the society is the ethic as strong as it is in the judiciary. In the one branch of the government that is not tied to the electoral process and sits astride the usual democratic system instead of fully within it, the ethic gathers added force from this uneasy status the judiciary holds. Yet, despite the strength of the ethic (or, more likely, because of it,) the judge in our legal system very often finds himself or herself in a very uncomfortable position, because law simply does not preexist to govern many (perhaps most) controversies. The "law" must be created, or invented, on the spot, which means that a judge who is conscientious about, and highly sensitive to, the need to abide by the ethic of restraining personal beliefs and values will frequently, if not continually, be faced with answering doubts about the real bases of the decisions the judge has arrived at. To minimize these doubts, the judge may naturally feel inclined to adopt a posture that provides the strong appearance of total personal disengagement or to follow an interpretive path that provides the least opportunity for interjection of personal values. At times both have occurred in free speech cases, and with unfortunate effects.

From the very outset of our discussion about the social functions of free speech, we have noted how uncharted has been the terrain open to the principle. The historical meaning of the amendment was unclear and, in any event, in all likelihood quickly exceeded; its language was general. How one should set about giving content to a "legal principle" in such a setting, particularly a constitutional principle, is a most complicated—and widely debated—issue.[11] Many would contend that there is a broadly shared, historically derived consensus about at least a core of meaning in the idea, a body of meaning external to the judges who have been charged with the task of interpreting the principle. Others, rejecting the possibility of a preexisting and identifiable public value of free speech, would emphasize the pure invention involved. Whatever position one takes in this debate over the methods of derivation of a free speech theory, there remains the crucial and significant reality for those actually engaged in giving content to the idea that the uncertainty of this task, when pitted against the ethic of restraining personal belief and of applying "the law," will likely generate behavioral impulses that end up distorting the analytical process involved, or at least its appearance.

This has happened with free speech in several identifiable ways. There is the type of rhetoric that implicitly, and sometimes explicitly, denies the truth of what is actually being done in the name of free speech: Here again we encounter the claims about the clarity of the historical meaning and language of the free speech clause, claims that are in fact unfounded; the characterizations of precedents as foreclosing choice when in truth they permit choice, and choice is being exercised; and the self-depictions of the court's actions as protecting the people against the government when in fact democratically reached decisions are being constitutionally overturned. All these we have encountered before in the discussion, but now we add the observation that such tendencies to misrepresent the reality of the judicial process in free speech cases may receive additional fueling from a judicial wish to appear choiceless in a context in which choices are being made and in which it is unclear what sources of value lie behind those choices.

There are other, less obvious ways in which the ethic of restraining personal beliefs in the exercise of the judicial function can exert a distorting effect on free speech rhetoric, ways that may be only an extension of those we have already considered but that are, in any event, worth singling out for special notice.

To begin, there is a noticeable tendency in our thinking, and particularly in judicial thinking, to isolate the significance of what is being done under the principle instead of noting its relevance to all social interaction. Concerned about the legitimacy of removing speech activity from social regulation, we respond by claiming, implicitly, that there is no harm in that removal, not only because the behavior sought to be regulated is inconsequential but also because we freely allow regulation of other behavior that really does matter. Here again is the characterization of speech acts as being only a prelude to action, with no real consequences worth taking account of. It also occurs, however, when speech is viewed as providing the opportunity or the justification for regulating nonspeech behavior.

When Mill noted the likely counterargument to his thesis that since we are not infallible, we must therefore tolerate all expression of ideas—the counterargument being that since fallibility is not a bar to regulation of nonspeech behavior, it ought not to be to regulation of speech—he answered by claiming that the regulation of speech was different: "There is the greatest difference between presuming an opinion to be true because, with every opportunity for contesting it, it has not been refuted, and assuming its truth for the purpose of not permitting its refutation. Complete liberty of contradicting and disproving our opinion is the very condition which justified us in assuming its truth for purposes of action; and on no other terms can a being with human faculties have any rational assurance of being right."[12]

Mill's response possesses strong appeal. All the time, we condition our acts that impinge on other people by permitting them to enter whatever objections they may care to raise; we think that to do so is only just and fair. In this sense, there is merit in the idea that the range of allowable speech activity is hinged to the fact that we choose to, and must, regulate other activity. But if we look at the whole matter from another angle (as we should, given the speech activity we actually tolerate), seeing in our responses to behavior generally a tendency to react improperly and undesirably as much with nonspeech as with speech behavior, and if we decide on that basis to "protect" speech against regulation, then it becomes important not to present the toleration of speech activity on the basis that it justifies the regulation elsewhere. To do so trivalizes the significance of the social meaning of the tolerance demanded under the free speech principle, because what we hope to learn by it is not just the moral lesson that fairness instructs us to listen to the arguments of our opponents, but rather that we must sometimes (perhaps even often) submit to their beliefs in order to live together peacefully. Examine any organization you like or, for that matter, any relationship: it is not listening alone that justifies the exercise of power, but the willingness to compromise and accommodate, too.

Reducing the meaning of speech toleration to the importance of listening also limits the symbolic value of the enterprise. Oddly enough, of all people, Mill was certainly one of the most keenly sensitive to the general tendency of every community to seek to impose an orthodoxy and to extirpate difference within itself, a problem not in the least bit limited to speech activities within the society; yet, rather than seeing the possibility of tolerance toward speech as a method of learning about that general tendency, he could see speech only as a way in which people voiced objections to other regulations. Thus, by isolating the meaning and significance of the toleration of speech, he not only too narrowly fixed the range of speech that could reasonably and beneficially be protected under a free speech principle but also risked losing the lessons of toleration for nonspeech behavior as well.

The isolation of speech harm (or of the needs of intolerance toward speech) as a tactic in trying to obtain greater adherence to the free speech principle undermines the very purpose of the free speech idea. It ought to be the purpose of free speech to see the connections between the problems that arise in our feelings toward speech and nonspeech behavior, and it obviously works against that objective to portray the treatment of speech as being somehow entirely "unique." When the point is in the similarities rather than in the differences, it makes little sense to rely on the differences. The potential loss from this way of talking, however, transcends that of a missed opportunity to make the pertinent point. It is possible that a consequence of this way of presenting free speech will actually accentuate the problems instead of alleviating them. While speech may be better secured as a result of the technique of isolation, nonspeech behavior may become *more* vulnerable to the impulse to excessive intolerance. Having satisfied ourselves by the toleration of speech activity that we are indeed a fair and self-restrained society, we may feel less need to be restrained or watchful toward our own behavior everywhere else. Like the man who feels freer to spend a dollar because he has just saved a dime, the tolerance toward speech can become an excuse for intolerance elsewhere.

Another tendency within free speech thought to limit the significance of the principle is to be found in the "process" vision of free speech. As we noted in chapter 5, the principle has come to be viewed as important for its vital systemic functions, for its role in feeding information and ideas into the democratic machine so that the best decisions can be made. The benefits identified as flowing from free speech are clear and specific and measurable: a better welfare system, a new method of solving crime, a better program for collecting taxes. The image we carry around with us is very much like the participants in Meiklejohn's town meeting or like a group of scientists interested in figuring out some truth about the external world: people who have agreed that they must be prepared to listen to all ideas because the truth may lie in any one or combination of them.

We have already considered both the serious limits of this vision for explaining what actually happens under the modern free speech principle and the principle's deeper levels or functions in affecting the nature of social interaction. Free speech is not by any means unrelated to the process by which social decisions are reached, but its actual relationship is more subtle and concerned with developing a general *capacity* rather than with feeding units of information into a mental machine.

The machine image is a powerful one for judges, partly because it seems to make it possible to distance themselves from value-laden choices. Taken to its extreme, the image turns judges into machines themselves, whose only function is to make sure that all the parts of the systemic machine are working properly. They are not called on to make judgments about whether the outcomes of the machine are good or bad, or whether the machine should be altered in this way or that so that the outcome will differ. The machine is as it is, for better or for worse, and the judge's only task is to let it do its work.

Such, for example, is the image of the judge under the First Amendment that so (momentarily, at least) attracted Alexander Bickel, who in his last book, *The Morality of Consent*, described the judicial function in process terms:

> Yet the First Amendment does not operate solely or even chiefly to foster the quest for truth, unless we take the view that truth is entirely a product of the marketplace and is definable as the perceptions of the majority of men, and not otherwise. The social interest that the First Amendment vindicates is rather, as Alexander Meiklejohn and Robert Bork have emphasized, the interest in the successful operation of the political process, so that the country may better be able to adopt the course of action that conforms to the wishes of the greatest number, whether or not it is wise or is founded in truth.[13]

This notion of the judge, as concerned only with ensuring that the "real" majority wins out, that governmental policy "conforms to the wishes of the greatest number, whether or not it is wise or is founded in truth," appeals to the desire of judges to restrain their own beliefs as they enforce the free speech principle. "I am not insisting that this speech be tolerated because I believe it is true or will produce truth," the judge can plead to those who wish to forbid the speech, "but because a *system* requires that the real majority rule, and that, in turn, requires that everyone know what the possibilities are. All I am doing, therefore, is making a process work, which all of you have agreed to. I have no grand designs for the society, no fundamental or metaphysical beliefs about the importance of truth or the best means to it, no goals for reforming the society whatsoever."

The desire of the judge essentially to remove himself or herself from any value involvement in free speech cases is very powerful, but it is both impossible to achieve and unfortunate from the standpoint of seeing and fully implementing the larger meaning of free speech, in particular, its relationship to the general capacity of tolerance. The judge simply cannot make the choices about which speech should be tolerated, and which not, without engaging in *some* value choices, whether the values be those of the judge or of someone else. Not only must you build a theory to explain why speech relating to the political system should be insulated from the operation of that very system, but you must also explain what you mean by, or will include under, the rubric "political speech." Does it include Cohen's inscription on his jacket? The Nazi speech in Skokie? Explicit advocacy of genocide? What will further majority choice, the model of totally unrestricted debate or the more restricted rules of the courtroom? These issues cannot be answered except by making reference to some additional principles, which require reference to additional values. Wellington's comment, quoted earlier, on the case against the result reached by the Court in *New York Times Co. v. Sullivan*, points up the unavoidably complex and difficult judgments involved in deciding to what degree legal protection will be afforded for protection against reputational injury. What rule will best advance a "democratic system of government" is far from clear and certainly not answered by declarations of intent to advance majority choice.

Quite apart from the possibility that judges might become mere administrative functionaries, if one believes it important and legitimate to have a free speech principle operating according to the tolerance theory, it then becomes a matter of some concern that judges may be dispositionally disinclined to accept that role, preferring instead to portray their role in more systemic and quantifiable terms. It is a curious irony that an excessive caution toward, or fear of, the impulses that are the concern behind the free speech idea can interfere with the effective operation of the idea itself, in this case arising from the institutional context created to implement the idea.

## III

A third source of pressure can distort thinking about free speech, in many of the same ways we have already identified with respect to the other pressures. Rather than arising from the belief that one is threatened by the desire to limit free speech or from the belief that judges must not inject their own convictions or values into their decisions, this pressure arises from the same need to dissociate oneself from the speech act being tolerated that those who seek censorship also have. This is every bit as complicated a business for judges as it is for those who object to the speech, but it is an added complication that makes the writing of judicial opinions especially difficult.

In drafting opinions, judges must decide to what extent they will indicate their own attitudes toward the speech in question. This is not so simple a matter as it might at first appear. In chapter 1 we noted that the judges in the *Skokie* opinions had stated their own feelings of personal abhorrence for the messages they were insisting had to be tolerated, and we also noted that this could constitute a form of official censure. It may also be recalled that Holmes in *Abrams* expressed his personal opinion that the defendants' views were "silly" and a "creed of ignorance and immaturity" and that the defendants themselves were only "poor and puny anonymities."[14] Similarly, Harlan, in *Cohen*, indicated that he thought Cohen's act was "a trifling and annoying instance of individual distasteful abuse of a privilege."[15] Should judges feel free to engage in such personal declarations? They should, though only with caution.

In at least three separate ways, such personal statements are critically important to the very operation of free speech, at least as it relates to increasing the capacity of social tolerance. First, it may well be that without some means of making a public declaration of personal feelings, judges would be psychologically unable to perform their required tasks. Since the official opinion is probably the only means of reaching a comparable audience, it is proper to extend judges access to this forum, so to speak.

Second, such personal statements may also be important in enhancing the function of the tolerance principle. If the judges actually agree with the speaker's views, the insistence on self-restraint by those who would censor the speaker will probably have less impact, since, of course, it is easy to be tolerant of that which you like. On the other hand, as we saw in chapter 5, what gives so many of these famous free speech opinions their power is precisely the tension between the judge's and the speaker's views, which parallels that between the community's and the speaker's views. In that setting the judges are able to develop a more keenly felt and a more persuasive claim for the benefits of toleration.

This brings us to the third point. In many of these cases, and certainly *Skokie* was one, the clash of arguments over tolerance and censorship is implicitly concerned with deeper issues, not just, or even necessarily, between the speaker and those who would censor but between those who would censor and the rest of the society. Tolerance may signify not only insensitivity to that segment of the society most deeply injured by the speech but also a wish to see that segment of people injured. Tolerance for the wrong reasons can be a form of vicarious aggression.

Several years ago a student of mine related that she had been assigned to attend a formerly all-white high school, under a general desegregation plan. Friction between the white and the black students arose almost immediately. Sometime during the year the student government decided to invite a major official of the Ku Klux Klan to speak at the school. When black students objected to the invitation, the white students responded that they were just in favor of free speech for everyone and against censorship, and that, while not agreeing with what the invited speaker was expected to say, they believed it was important to hear every side of every issue. Understandably, the black students felt that beneath this mask of commitment to open-mindedness was an act of unmistakable hostility.

We have already observed how this same fear over what really lay at the root of the demand for toleration was present in the

*Skokie* case. Lurking always just beneath the surface was a question of whether the free speech position of many was really premised on anti-Semitic prejudice that could be conveniently implemented in this covert way.

"Free speech," in other words, can be many things, including a means of inflicting injury. To be that, moreover, it need not be so overt, or direct, as it was in the instance involving my former student, or as some may have feared in *Skokie*. The hostility may be manifested simply by declaring a desire to "consider" viewpoints others find objectionable, a phenomenon all parents of adolescent children must feel they know all too intimately. It is, therefore, neither surprising nor undesirable that judges in these cases should take it upon themselves to express their "personal" sentiments, and even perhaps what they view to be the nation's sentiments, toward the speech being tolerated. It is important to allay the fears of some that they are being victimized in the name of freedom of expression and to allay the concerns of others that self-restraint signifies tacit approval. But the fact remains that a form of censure is present in this act of personal declaration, and judges should accordingly view it as an important step, which should be taken only when the considerations we have just identified fairly merit it.

## IV

If free speech is to symbolize a commitment to developing controls over a general impulse to intolerance, by focusing on that impulse in one sphere of human interaction, it is supremely important that the principle be *implemented* with sensitivity to the various pressures that distort human thought in that context as well as others. In the foregoing discussion we looked at three such potential pressures, which together or independently exert a kind of magnetic attraction on free speech thinking, deflecting it from the course it ought to be taking. Some distortions may be the product of any one or of all three sources: The false portrayal of difficult cases as settled by the clarity of the First

Amendment text, for example, may be the result of an excessive fear of considering arguments for censorship in this particular instance or any other, or the wish of the judge to appear as not implementing what he or she happens to think is best for the society, or the need of the judge to indicate that the toleration insisted on does not arise from any personal sympathy with the viewpoints expressed. These suggestions about various contextual needs that may distort the presentation of free speech, however, are most important for what they illustrate about the kind of inquiry we should undertake to a greater degree than has been done. The problems we face in administering the free speech principle go beyond those of faulty logic.

We ought to think of free speech, at least partly, as an institutional forum in which certain issues are addressed. The primary, though by no means the exclusive, speech is that given by the judges at the conclusion of the case. We compel judges to speak for many reasons: the necessity to explain is thought to induce honesty and thoughtfulness, as well as to offer an invitation to others to engage in dialogue. But, as Meiklejohn acutely observed, the judicial opinion is more than a means of simply identifying the result reached, it is also—and perhaps most significantly—a forum for education. It is a "public corner" on which those who have been assigned the task of settling disputes over speech activity must stand up and speak out on the controversy. In doing so, they must remind us of the reasons for having the principle in the first place, as well as explain how they applied the principle to the facts of the particular case. How they do this will illustrate for us the capacity for intellectual character that we should be seeking. It seems a reasonable assumption that if they are engaged in weaving a set of deceptions, of censoring reality for themselves and for others, neither they nor we will achieve the goal.

At the very least, seeing the problems and difficulties involved in the demand for censorship of speech will provide us with a foundation on which to engage in self-examination of the ways in which we approach and deal with conflicts over speech activity—

with how we think about speech activity itself, with how we deal with those who demand the right to censor, and with how we struggle with the role of judicial review. In short, the same theory of free speech that instructs us on the ends for which the principle is aimed, and provides an aid for its application, also provides us with a source for self-criticism, a set of premises about human thought and action that applies every bit as much to free speech disputes as to any other social controversy.

# 8

# An Agenda for the General Tolerance Theory

Wigmore laid the foundations for a powerful critique of modern thinking about the free speech principle. There is a distressing tendency in free speech thought and discourse to exaggerate the evils of government and the goodness of people, to minimize the value to society of the freedom to impose punishment on speech activity, and to understate the risks and harms of speech and to overstate its benefits. But Wigmore, like so many others, could not imagine that a free speech principle could function as other than a protection of valuable discourse or a preserve of personal liberty, perhaps with some room left at the margins to secure the inner core. Thus, to him the notion of tolerating extremist speech seemed anomalous, paradoxical, even perverse. Nothing in the fundamental theory of the principle appeared to sanction such an extension. In fact, that fundamental theory appeared to contradict it: A commitment to democracy, or to truth seeking generally, would seem to lead one to want to prohibit, not protect, speech activity designed to overturn those processes.

There are many paths by which a society, just as an individual, may seek to form itself into the type of community it aspires to be. The most common course is to restrict and punish undesirable and unwanted behavior. In doing so, by segregating certain behavior for condemnation and punishment, the community affirms for itself the correct way of being. An alternative, which is nearly as common as proscription, is to reward good behavior. Here again the societal response bespeaks a general affirmation of, or commitment to, a way of being, though it is by the opposite route of applauding and celebrating the most sought after behavior.

With these two methods, proscription and reward, we are familiar and comfortable. We encounter them at work virtually wherever there is social interaction.

Other methods are available, however, to the society that seeks to control and channel the impulses and capacities of its members. In free speech we find one of the most significant: the toleration of undesirable and unwanted behavior as a method of pointing up troublesome tendencies within those wishing to be intolerant, often by the community's engaging in self-restraint toward the very behavior it seeks to avoid. By this vision, free speech bears the burden not simply of correcting a deficiency perceived to be present in the context of legal censorship of speech but one much more widely present—a shared general characteristic. The problem addressed is universal in a longitudinal and latitudinal sense: everyone knows it and knows it in every context. Addressing such a broad issue, free speech stands symbolically as the gateway to social intercourse.

## I

Why free speech should have evolved into this meaning at this particular time in American history is, of course, a very difficult question. It seems likely that the ordering of the several virtues will vary from society to society, depending on the conditions prevailing. For a culture threatened by external aggression or bent on conquest, courage and honor will be the most prized. For a country like the United States, tolerance appears to have assumed a leading position. We have observed in the course of the discussion a number of ways in which the containment of belief, and the impulses associated with it, and of the fears at the thought processes of others, serves a variety of important social functions. To these might be added the consideration that a capitalist economic system requires a broad capacity for self-containment. A basic ethic of our public corporation law (though in the last two decades it has come under considerable attack) is that managers of public corporations must not employ their power to implement their personal values or social preferences.[1] Additionally, it may be thought that a free enterprise economic structure, which certainly bears the appearance of wide disorder, requires a considerable capacity to accept a highly fractionalized social system. It is perhaps no accident that many of the major metaphors used in the articulation of the modern free speech principle are drawn from the free enterprise lexicon: we hear of the "free trade in ideas" and of the "marketplace of ideas."

As we have also seen, a strong capacity for tolerance is also required for a society with pervasive bureaucratic and professional systems. The performance of tasks within each of these sectors of society requires the ability to submerge the self. One's own values are not to intrude into one's performance. Here again, of course, the judge is the prototype for this capacity.

It is sometimes suggested in discussions about the rise of free speech in American society that such an idea, in some form, was absolutely essential in a social setting with an extremely high percentage of immigrant populations from divergent cultures and with conflicting ways of life and values. The potential for conflict inherent in this intermix of cultures was so great that some policy of mutual toleration had to be devised—at least as long as no single group or alliance could hope to acquire a firm grip on the reins of power within the society. "[I]f you have two religions in your midst they will cut each other's throats; if you have thirty, they will live in peace," said Voltaire of tolerance.[2] While there is undeniable truth to this observation about the need for mutual toleration in a society of diverse groups, it does not seem to provide a wholly satisfying explanation for why highly extremist and inflammatory speech should have been tolerated instead of punished. In fact, the same premise might be thought to compel precisely the opposite result.

Perhaps a contrary thesis is more true, one that would emphasize the enormous range of similarity, of shared values cutting through the society, rather than the differences. Perhaps only with the development of a highly stable society, in which people widely share a relatively common set of values, could a real capacity for toleration of virtually unrestricted speech activity emerge. American politics, with its homogeneous two-party system (homogeneous in comparison with the splintered political life of most European societies), both undergirds and symbolizes this unified perspective throughout the country. Since, as a practical matter, deviant groups have such a small chance of supplanting or making serious inroads into the existing political hierarchy, it is a matter of far less moment (and actually some considerable advantage, as the discussion in chapter 5 revealed) that they be tolerated.

The evolution of our attitudes toward toleration deserves further study, and a part of that investigation should be to try to account for why its newer functions have not been fully recognized. We have already considered several possible answers to this question, but it is worth reemphasizing one—that a social practice or principle like free speech often can perform several functions simultaneously. Because it has accretions of meanings, not just one meaning at a time overtaking that which went before it, the newer functions may be temporarily overlooked. The language and thought patterns of the traditional functions may run counter to those of the new ones—as the carefully wrought mental separation of "word" and "deed" may have obscured the tolerance principle.

It may be important to recall that the traditional concept of "liberty," as a zone of behavior beyond the reach of the state, is itself a relatively recent notion, extending back only a few centuries. As the political system gradually moved from autocracy to democracy, it was natural for speech activity to be regarded as an integral and a vital element to that process. These meanings, moreover, continue into the present. There has never been, and probably will never be, a sense of complete identification between the citizenry and the government; some antagonism, some sense of estrangement, between the two seems likely to be an inexorable reality. The threat of the government's abusing its delegated powers remains a real, if reduced, risk, one increased by public apathy (which may in turn at least partially explain the frequent critical characterizations of government one finds in free speech opinions—making free speech jurisprudence a form of sedition for democratic ends).[3] Finally, it would be naive to think there is, or is soon likely to be, a total mutual trust prevailing among the citizens themselves. The phenomenon of the tyranny of the majority, though perhaps playing a more latent role in today's society, remains a factor and is often reflected in a general wariness one occasionally senses between groups within the nation.

In short, what we think of as the original meanings behind the development of the free speech principle continue to have contemporary significance, and that reality may help account for why newer functions may be overlooked: historical meanings can screen attention from newer ones. We can see this in operation with retrospectives on censorship, like that held just a few years ago at the New York City Public Library. Our attention is focused on the most horrendous examples of suppression of speech, acts of expression we now regard as possessed of great value—like the censorship of Joyce's *Ulysses* or of Lawrence's *Lady Chatterley's Lover*. We think, and properly so, that such acts of censorship reflected gross insensitivity to acts of expression that civilization values most, but we conclude our thinking with a grateful genuflection to the principle of freedom of speech, which puts a curb on such misbehavior.

In doing this we associate freedom of speech only with the objective of preserving speech of great merit, or perhaps of preserving an area of freedom for each of us that is beyond the reach of the state. While worthy goals, they lead us to overlook the fact that the feelings that generated the instances of gross censorship we deplore play a significant role, both good and bad, in guiding virtually every aspect of social interaction. Rather than praising free speech for its protection of works of great merit, and then being forced under the pressure of an extremist speech case to turn to defend the constitutional principle on the supposedly "neutral" basis that it simply protects a "process" from legal intervention, we might instead renew our understanding of the principle by taking a sympathetic look at the feelings behind that intolerance. If we did, in the end we will retain a view of free speech as having a moral dimension, of being one public context in which the society addresses an important aspect of the general quality of mind it seeks.

## II

Looking at it from a distance, therefore, we may more readily see the potential for an evolution of meaning with an enterprise like freedom of speech. Different circumstances may lead people to focus on different reasons for having the principle. One generation may seek merely the right to speak, while another will take strength from tolerating bad speech acts. Each generation has its own agenda, as the conditions then prevailing will dictate what subjects are important for inquiry. In a letter to his wife John Adams once noted that it was his task in life to "study politics and war." If successful at those subjects, he said, then his children would be free to "study mathematics and philosophy, geography, natural history and naval architecture, navigation, commerce and agriculture." And they, if successful at these, would "give their children a right to study painting, poetry, music, architecture, statuary, tapestry and porcelain."[4] So it can be with free speech, as the constraints and circumstances of the period will determine what we draw from the experience it offers.

It is common to hear people speak of this society's commitment to what is called pluralism, by which it is generally meant that we are a people prepared to accept a diversity of viewpoints and life-styles. Insofar as this refers to a general idea of a presumption of tolerance, or what I have at times referred to as the tolerance ethic, it is consistent in a very broad way with the view of the underlying purpose of free speech presented here. But whatever one may take to be the advantages of pluralism—as a stimulus for finding truth or a source of pleasure because of the variety of life-styles encountered or a source of compromise needed for a stable society—it will not do to consider free speech simply an ordinary example of such a commitment. The principle goes too far to be explained in terms of these traditional notions of pluralism. It is not persuasive to argue that Nazis should be permitted to march in a Jewish community because we believe in a pluralistic society. Even pluralism must have its limits. Much more must be shown before we should give our assent to such a result.

But it is possible to say the following: that the commitment to developing a capacity to accept diversity in broad areas of social interaction is desirable, for a number of reasons we can offer; that we must, however, be prepared to draw limits on our acceptance of diversity—we must decide when we have had enough of pluralism and can accept no more; that drawing these lines requires a capacity for good judgment, which cannot be reduced to a simple rule, or set of rules, because of the variety and complexity of the contexts in which the decisions have to be made; that because of the sheer impossibility of formulating in advance a rule for decision, it is helpful if we become conscious of those elements in our thinking that tend to skew and distort our judgments; and, finally, that freedom of speech is a context in which those general biases arise and can efficiently be highlighted through a practice of extraordinary self-restraint.

## III

But can such a principle work? What assumptions about the society must be made in order for us to accede to the principle?

What should we be investigating as we now consider the principle in the light of the general tolerance theory?

The purpose of the preceding chapters has been to argue that the constitutional principle of free speech has taken on important new meaning in this century. I have tried to describe that general function and to articulate its rationale. The new principle has been presented as an appealing and intriguing social response to a critically important social issue. But there are also good grounds for reservations about the principle. All the components of the theory—the assumptions about the nature and degree of the impulse to intolerance, about the advantages of using speech as a discrete area in which to engage in extraordinary self-restraint toward troublesome behavior, about the capacity of the judicial system to effectively implement such a principle, about the role of law in shaping social attitudes generally, and others—are matters that, quite obviously, may be challenged as unworkable in practice, even if sound in theory.

One broad area of concern with the free speech principle has not been raised but should be noted: We ought to consider whether the actual operation of free speech might have the ironic result of stimulating excessive intolerance elsewhere in the society.

The possibility that free speech might do this can be examined on at least two levels. We may begin by considering whether the identified impulse to excessive intolerance is really so entrenched in human nature as to be entirely resistant to any form of treatment, at least any treatment such as we have in mind with respect to the new free speech principle. Perhaps the impulse is like a pillow, and free speech is just a strike at the pillow. In that case, when one part is punched in, the whole only redistributes itself into a new shape and form; the total volume remains the same, despite a new indentation in one area of its surface. Could this be true with efforts to address the impulse to excessive intolerance through the indentation of free speech?

Even if the impulse can be effectively moderated, will the free speech enterprise actually work to achieve that moderation?

Here we draw once more on that curious phenomenon in the free speech area of how both sides, those who are proponents of free speech and those who favor suppression, draw ultimately on the same stable of arguments for making their cases. Now we see how a traditional free speech claim can be turned the other way: Might it not be the case, it can be asked, that by prohibiting intolerance we will only end up making it the *more* attractive to people in other social contexts?

The matter is complicated. In chapter 7, when we discussed the tendency among judges and free speech theorists to think about and present free speech in such a way as to isolate it from having any broader meaning, it was assumed that the tendency could be corrected. But what if it can not? What if those who implement free speech will invariably be predisposed to treat it as an isolated phenomenon? Then, of course, the risk is, or will be, that the identity of "tolerance" achieved through free speech will become an *excuse* for giving vent to excessive intolerance in other areas.

From still another side it may reasonably be asked whether free speech is working on the wrong problem or will inevitably go too far in its quest for building a greater capacity for tolerance. Free speech may be *too* successful and create a problem of excessive tolerance, which may have equally pernicious effects for the society as its opposing vice. We have already noted that tolerance and intolerance are not ends of a spectrum of good and bad, even in the context of legal coercion against speech, despite the frequent portrayal of them as essentially moral opposites. The human condition, however unfortunately, is far more difficult; it is, as Aristotle said, a problem of finding the right balance in every context. Free speech thought has always placed a decided emphasis on seeing the pole of intolerance as posing a much greater social risk than does that of tolerance, and, acting on that premise, we have erected an elaborate and sophisticated system of legal doctrine. But that is an assumption worth reexamining every now and then, because it may be a truer estimate of social life to think that the more likely deviation from that difficult middle course of virtue varies over time, and that what Dr. Johnson said about the greater difficulty of supplying a deficiency, rather than an excess, of action can sometimes be true of excessive toleration as well:

> It may be laid down as an axiom, that it is more easy to take away superfluities than to supply defects; and, therefore, he that is culpable, because he has passed the middle point of virtue, is always accounted a fairer object of hope, than he who fails by falling short. The one has all that perfection requires, and more, but the excess may be easily retrenched; the other wants the qualities requisite to excellence, and who can tell how he shall obtain them? We are certain that the horse may be taught to keep pace with his fellows, whose fault is that he leaves them behind. We know that a few strokes of the axe will lop a cedar; but what arts of cultivation can elevate a shrub?[5]

With free speech we must always beware lest we end up seeking to shore up our defenses on the side that needs it the least, while the natural barriers in the opposite direction, though they now appear to us the more secure, are in the larger scale of human events the more fragile and vulnerable. We would then be, once again, creating a Maginot Line against the wrong threat.

The pursuit of toleration may be complicated in another sense. At some point, tolerance seems to blend into what we think of as obedience. In a way, the soldier is the most tolerant of individuals. Holmes, in fact, sometimes wrote of how to him the most admirable mind was that of the obedient soldier—a not unexpected logical extension of Holmes's efforts to eradicate belief while retaining the desire to play a social role.[6] The doctrine of papal infallibility demands considerable "tolerance" of the Catholic laity and clergy. And the old secular version of this idea, that the king spoke with divine authority and could do no wrong, also demanded considerable capacity of toleration from the subjects. Yet, quite clearly, this seems not to be all that we have in mind with the free speech principle. But might that be all it will accomplish?

We must become comfortable with the idea that free speech

is concerned with the development of a mind that is itself comfortable with uncertainty and complexity. The obedient mind does not think for itself. It may have beliefs, which may be inconsistent with the orders received from above, but those beliefs must be thoroughly suppressed. The individual must be instructed to disregard, even disrespect, his or her own mind.

The mind sought through free speech is distinguishable from the obedient mind in this one critical respect: While possessing an element of this capacity for self-criticism and doubt, as well as for other ways of controlling belief, the tolerant mind sought through free speech is free to consider openly, to entertain seriously, the possibility of disobedience. The object is not to hide reality from oneself, to "censor," in the broadest sense of that term. The obedient mind functions through an act of stern censorship, the tolerant mind of free speech does not. We must think of free speech as being concerned only with a *bias* in the direction of belief, or with the desire to believe, not with a total silencing of any independent thought.

But, still, the risk is that free speech will become a method of inculcating a kind of toleration that turns naturally into passivity and uncritical obedience. While there is, as was argued in chapter 7, a definite and sometimes very great social need, in some contexts and on some occasions, for a noncritical, nonindependent mind, the aim of free speech is to work generally against that kind of intellectual character.

One final item should be included on the future agenda: It would be desirable if free speech scholarship were to turn somewhat away from the traditional problems it has addressed, which involve primarily the intricacies of the tests employed in the various areas of First Amendment litigation, and examine more broadly what the impact of the concept seems to be in social thought generally. Free speech is too vital a national symbol to be thought about exclusively in doctrinal terms. Devising a test for deciding who is a "public figure" for purposes of the constitutional standards regarding libel, or what is "obscene" for purposes of the exception for pornography, is unquestionably important. But we also need to look at how the concept of free speech affects social decision making *beyond* the ken of legal restraints on speech, because that larger connection appears to fuel the meaning, or at least a significant part of the meaning, of the principle itself. When we hear people speak of free speech as embodying ideas like "having respect for other people's opinions," we know that the legal principle is functioning as part of a general social ethic and not simply as a means of curtailing legal intervention into the realm of speech. That connection merits further study.

## IV

At the outset of this book, we wondered whether free speech had come to reflect an excessively single-minded quest for a one-sided view of life or a kind of outmoded, antiquated idea better suited to the political conditions of an earlier era. The remainder of the book has tried to show that, rather than being one-sided or anachronistic, free speech still plays a critical role in an integrated social life. It can play that role even though liberty of speech, in any of the classical meanings given to that notion, may reasonably be regarded as having been realized. The reason behind that continued meaning lies in the nature of what is revealed in the *response* to speech activity, something that often manifests itself in graphic form in that context but arises elsewhere, too.

When we see this, our perspective on the social function of free speech shifts dramatically. Our attention is turned from the value of speech to the troublesomeness of the reactions evoked by speech. Extremes become integral to the social meaning of the idea, and the broad connections with other areas of social interaction are highlighted.

In the end, we must not fail to see the genuine nobility of a society that can count among its strengths a consciousness of its own weaknesses.

# Notes

## Chapter 1

1. Collin v. Smith, 578 F.2d 1197 (7th Cir.), *cert. denied*, 439 U.S. 915 (1978); Village of Skokie v. National Socialist Party of America, 69 Ill.2d 605, 373 N.E.2d 21 (1978).

2. Aryeh Neier, *Defending My Enemy* (New York: Dutton, 1979), 79.

3. 250 U.S. 616, 624 (1919) (Holmes, J., dissenting).

4. The Espionage Act was first enacted by Congress in 1917. The act made it a criminal offense to (among other things) "willfully make or convey false reports or false statements with intent to interfere with the operation or success of the military or naval forces of the United States or to promote the success of its enemies, and ... [to] willfully cause or attempt to cause insubordination, disloyalty, mutiny, or refusal of duty, in the military or naval forces of the United States, or ... [to] willfully obstruct the recruiting or enlistment service of the United States ...." In 1918 Congress amended the act to forbid a variety of antiwar activities; those important to the *Abrams* case (250 U.S. at 616–17) are recited in the Court's description of the indictment:

> On a single indictment, containing four counts, the five plaintiffs in error, hereinafter designated the defendants, were convicted of conspiring to violate provisions of the Espionage Act of Congress (§3, Title I, of Act approved June 15, 1917, as amended May 16, 1918, 40 Stat. 553).
>
> Each of the first three counts charged the defendants with conspiring, when the United States was at war with the Imperial Government of Germany, to unlawfully utter, print, write and publish: In the first count, "disloyal, scurrilous and abusive language about the form of Government of the United States"; in the second count, language "intended to bring the form of Government of the United States into contempt, scorn, contumely and

disrepute"; and in the third count, language "intended to incite, provoke and encourage resistance to the United States in said war." The charge in the fourth count was that the defendants conspired "when the United States was at war with the Imperial German Government... unlawfully and wilfully, by utterance, writing, printing and publication, to urge, incite and advocate curtailment of production of things and products, to wit, ordnance and ammunition, necessary and essential to the prosecution of the war."

5. 249 U.S. 47 (1919). The other cases constituting the trilogy were Frohwerk v. United States, 249 U.S. 204 (1919), and Debs v. United States, 249 U.S. 211 (1919). The defendants in these cases were prosecuted under the Espionage Act as originally enacted in 1917. See supra n. 4.

6. 249 U.S. at 51.

7. Id. at 52.

8. Id.

9. Id.

10. Id.

11. Id.

12. 250 U.S. at 627, 628.

13. Id. at 628.

14. Id. at 629.

15. Id.

16. Id. at 630.

17. Id. at 631.

18. See, e.g., Dennis v. United States, 341 U.S. 494 (1951).

19. 250 U.S. at 630.

20. See Wigmore, "Abrams *v.* U.S.: Freedom of Speech and Freedom of Thuggery in War-Time and Peace-Time," 14 Ill. L. Rev. 539 (1920).

21. Id. at 549.

22. Id. A similar charge was made by the nineteenth-century antagonist of John Stuart Mill, the English barrister James Fitzjames Stephen, who said of the terms *liberty, equality*, and *fraternity* that "in the present day even those who use those words most rationally... have a great disposition to exaggerate their advantages and to deny the existence, or at least to underrate the importance, of their disadvantages." J. F. Stephen, *Liberty, Equality, Fraternity* (Cambridge: Cambridge University Press, 1967), 53.

23. 250 U.S. at 628.

24. Supra n. 20, at 550.

25. Id. at 558.

26. Id. at 557–58.

27. Id. at 559.

28. Id. at 554 (emphasis omitted).

29. Id.

30. Isaiah Berlin, "Does Political Theory Still Exist?" in Peter Laslett and Walter Garrison Runciman, *Philosophy, Politics, and Society*, 2d ser. (Oxford: Basil Blackwell, 1962), 19.

31. 578 F.2d at 1199 n. 1.

32. The ordinances are reproduced at n. 47, infra.

33. Smith v. Collin, 439 U.S. 916 (1978). The two dissenters, Blackmun and White, would have granted *certiorari* in part because of the "pervading sensitivity of the litigation.... On the one hand, we have precious First Amendment rights vigorously asserted and an obvious concern that, if those asserted rights are not recognized, the precedent of a 'hard' case might offer a justification for repression in the future. On the other hand, we are presented with evidence of a potentially explosive and dangerous situation, inflamed by unforgettable recollections of traumatic experiences in the second world conflict." Id. at 918.

34. See Neier, *Defending My Enemy*, 40.

35. Id. at 17.

36. See *N.Y. Times*, 10 July, 1978, 14, col. l.

37. 578 F.2d at 1200.

38. Id.

39. Id. at 12lo.

40. 69 Ill.2d at 612, 619.

41. 578 F.2d at 1200.

42. Id. at 1201.

43. Id.

44. 315 U.S. 568 (1942).

45. Id. at 571, 572. The passage, in its entirety, reads:

> There are certain well-defined and narrowly limited classes of speech, the prevention and punishment of which have never been thought to raise any constitutional problem. These include the lewd and obscene, the profane, the libelous, and the insulting or "fighting" words—those which by their very utterance inflict injury or tend to incite an immediate breach of the peace. It has been well observed that such utterances are no essential part of any exposition of ideas, and are of such slight social value as a step to truth that any benefit that may be derived from them is clearly outweighed by the social interest in order and morality.

46. 343 U.S. 250 (1952).

47. The Illinois statute in *Beauharnais* provided:

> It shall be unlawful for any person, firm or corporation to manufacture, sell, or offer for sale, advertise or publish, present or exhibit in any public place in this state any lithograph, moving picture, play, drama or sketch, which publication or exhibition portrays depravity, criminality, unchastity, or lack of virtue of a class of citizens, or any race, color, creed or religion which said publication or exhibition exposes the citizens of any race, color, creed or religion to contempt, derision, or obloquy or which is productive of breach of the peace or riots....

Illinois Criminal Code, Ill. Rev. Stat., 224a, 1949, c.38, Div. 1, 471. Relevant portions of the Skokie ordinance provided as follows:

> SKOKIE VILLAGE ORDINANCE No. 77–5-N–994 "An Ordinance Relating to Parades and Public Assemblies"
>
> Section 27–51. Permit Required. No parade, public assembly or similar activity,

where the number of participants expected may reasonably be assumed to exceed fifty (50) or more persons, and/or vehicles is permitted on any street or area of the Village unless a permit allowing such activity has been obtained from the Village Manager, or upon referral, to the President and Board of Trustees; provided, however, that this Ordinance shall not apply to students going to and from classes, or participation in educational activities under the immediate direction and supervision of school authorities or a governmental agency acting within the scope of its functions, nor shall a permit be required for normal or scheduled activities of the Village.

Section 27–54. No permit shall be issued to any applicant until such applicant procures Public Liability Insurance in an amount of not less than Three Hundred Thousand Dollars ($300,000) and Property Damage Insurance of not less than Fifty Thousand Dollars ($50,000). Prior to the issuance of the permit, certificates of such insurance must be submitted to the Village Manager for verification that the company issuing such insurance is authorized to do business and write policies of insurance in the State of Illinois.

Section 27–56. Standards for Issuance.
*b.* The activity will not create an imminent danger of a substantial breach of the peace, riot, or similar disorder.
*c.* The conduct of the parade, public assembly, or similar activity will not portray criminality, depravity or lack of virtue in, or incite violence, hatred, abuse or hostility toward a person or group of persons by reason of reference to religious, racial, ethnic, national or regional affiliation.

SKOKIE VILLAGE ORDINANCE No. 77–5-N–995 "An Ordinance Prohibiting the Dissemination of Materials Which Promote and Incite Group Hatred"

Section 28.43. Prohibition of materials promoting or inciting group hatred.

Section 28.43.1. The dissemination of any materials within the Village of Skokie which promotes and incites hatred against persons by reason of their race, national origin, or religion, and is intended to do so, is hereby prohibited.

Section 28.43.2. The phrase "dissemination of materials" includes but is not limited to publication or display or distribution of posters, signs, handbills, or writings and public display of markings and clothing of symbolic significance.

SKOKIE VILLAGE ORDINANCE No. 77–5-N–996 "An Ordinance Prohibiting Demonstrations by Members of Political Parties Wearing Military-Style Uniforms"

Section 28.42.1. No person shall engage in any march, walk or public demonstration as a member or on behalf of any political party while wearing a military-style uniform.

Section 28.42.2. "Political party" is hereby defined as an organization existing primarily to influence and deal with the structure or affairs of government, politics or the state. . . .

48. See, e.g., Miller v. California, 413 U.S. 15 (1973); Roth v. United States, 354 U.S. 476 (1957).

49. See "Testimony On The Psychological Effects Of Racial Slurs," by psy-

chiatrist David Gutman, Defendant's Exhibit 13; Collin v. Smith, 447 F. Supp. 676 (N.D. Ill., E.D. 1978).

50. See, e.g., Young v. American Mini Theaters, Inc., 427 U.S. 50 (1976); Kovacs v. Cooper, 336 U.S. 77 (1949); Adderly v. Florida, 385 U.S. 39 (1966).

51. 578 F.2d at 1203. The court's finding on this point was apparently based on Supreme Court rulings qualifying *Chaplinsky* in this respect, namely, Gooding v. Wilson, 405 U.S. 518 (1972), and Cohen v. California, 403 U.S. 15 (1971).

52. Gertz v. Robert Welch, Inc., 418 U.S. 323, 339 (1974), quoted in *Collin* at 1203.

53. See 578 F.2d at 1204.

54. See id. at 1202.

55. 395 U.S. 444.

56. Brandenburg, at the meeting, said, "The Klan has more members in the state of Ohio than does any other organization. We're not a revengent [*sic*] organization, but if our President, our Congress, our Supreme Court, continues to suppress the white, Caucasian race, it's possible that there might have to be some revengeance [*sic*] taken. . . ." 395 U.S. at 446. Some at the meeting carried weapons. Brandenburg was convicted under the Ohio Criminal Syndicalism statute for advocating unlawful activity and for assembling with a group formed to encourage "the doctrines of criminal syndicalism."

57. See 578 F.2d at 1204.

58. Laurence H. Tribe, *American Constitutional Law* (Mineola, N.Y.: Foundation Press, 1978), 689. The best modern analysis of the concept of "content regulation" is Stone, "Content Regulation and the First Amendment," 25 Wm. & Mary L. Rev. 189 (1983).

59. See 578 F.2d at 1201–2.

60. 403 U.S. 15 (1971).

61. Id. at 21.

62. Id. at 25.

63. 578 F.2d at 1210.

64. Id. at 1201.

65. The standard work on legal fictions is Lon Fuller, *Legal Fictions* (Stanford, Calif.: Stanford University Press, 1967). I have discussed the role of the line-drawing claim in legal argumentation in "The Sedition of Free Speech," 81 Mich. L. Rev. 867 (1983).

66. For materials generally on the conventions (including copies of the conventions themselves) the reader should consult Louis B. Sohn and Thomas Burgenthal, *International Protection of Human Rights* (New York: Bobbs-Merrill, 1973), and its companion volume, Louis B. Sohn and Thomas Burgenthal, *Basic Documents on International Protection of Human Rights* (New York: Bobbs-Merrill, 1973).

Article 4 of the Convention on the Elimination of All Forms of Racial Discrimination contains extensive prohibitions against racist speech activities:

> States Parties condemn all propaganda and all organizations which are based on ideas or theories of superiority of one race or group of persons of one color or ethnic origin, or which attempt to justify or promote racial hatred and discrimination in any form, and undertake to adopt immediate and

positive measures designed to eradicate all incitement to, or acts of, such discrimination and, to this end, with due regard to the principles embodied in the Universal Declaration of Human Rights and the rights expressly set forth in Article 5 of this Convention, inter alia:

(*a*) Shall declare an offense punishable by law all dissemination of ideas based on racial superiority or hatred, incitement to racial discrimination, as well as all acts of violence or incitement to racial discrimination, as well as all acts of violence or incitement to such acts against any race or group of persons of another color or ethnic origin, and also the provision of any assistance to racist activities, including the financing thereof;

(*b*) Shall declare illegal and prohibit organizations, and also organized and all other propaganda activities, which promote and incite racial discrimination, and shall recognize participation in such organizations or activities as an offense punishable by law;

(*c*) Shall not permit public authorities or public institutions, national or local, to promote or incite racial discrimination.

The convention (like the Convention on the Prevention and Punishment of the Crime of Genocide) has been ratified by dozens of countries, including Australia, Austria, Canada, China, Denmark, France, the Federal Republic of Germany, Greece, Iceland, India, The Netherlands, New Zealand, Norway, Spain, Sweden, Switzerland, and the United Kingdom.

When the United States signed the convention, it issued the statement: "The Constitution of the United States contains provisions for the protection of individual rights, such as the right of free speech, and nothing in the Convention shall be deemed to require or to authorize legislation or other action by the United States of America incompatible with the provisions of the Constitution of the United States of America." L. Sohn and T. Burgenthal, *International Protection of Human Rights*, 861–62.

Various commentators have recommended that the United States not ratify the convention, or ratify it only with a reservation similar to the one made at the time of signing. See Richard B. Lillich and Frank C. Newman, *International Human Rights: Problems of Law and Policy* (Boston: Little, Brown, 1979), 135–57.

Rita Hauser, the U.S. Representative to the UN Human Rights Committee, doubted that the United States would ratify the convention: "The ideas in Article 4 fly directly against the American idea of free speech." Hauser, "United Nations Law on Racial Discrimination," 64 Am. J. Intl. L. 114, 117 (1970).

I should like to note here that, while I shall ultimately develop a theory for explaining why extremist speech ought to be protected under the First Amendment, I do not intend to speak to the question whether such an interpretation of the First Amendment should be taken as precluding the United States from ratifying either of the two conventions.

Most Western European countries have laws prohibiting incitement of racial hatred. In Great Britain, the Race Relations Act of 1965 created the offense. In 1976 the pertinent section was amended and inserted in the Public Order Act of 1936. As sec. 5A, "incitement to racial hatred," it now reads:

(I) A person commits an offense if—
(*a*) he publishes or distributes written matter which is threatening, abusive or insulting; or
(*b*) he uses in any public place or at any public meeting words which are threatening, abusive or insulting.

If convicted of incitement to racial hatred, the offender may be fined or imprisoned up to two years. Public Order Act 1936, sec. 5A (5)(*b*).

France and Denmark have similar laws making it an offense to incite racial hatred. In France the convention is part of French law. Racial defamation, racial insults, and incitement to racial discrimination, hatred, or violence are misdemeanors, punishable by imprisonment for one month to one year. Art. 1, 3, Act 72–546, of 1 July 1972.

The 1972 law was first applied to the publisher of the magazine *Bulletin U.R.S.S.* An article published in the magazine said that Jews are required by their religion to kill non-Jews and to refuse to help non-Jews. The publisher was convicted of racial defamation and incitement to racial hatred. He was fined 1,500 francs and directed to publish the court's decision in his magazine and six newspapers. Tribunal correctionnel de Paris, 24 April 1973, Jurisprudence française 1974, 529.

Denmark's law on racist speech predated the convention, but it amended the law after ratifying the convention. Art. 2666 provides for punishment for public statements threatening, insulting, or degrading because of race, color, national or ethnic origin, or belief.

The Italian law is a modified enactment of the convention, punishing individual incitement to hatred as well as prohibiting associations that have the goal of incitement to racial hatred. An individual may be imprisoned for up to four years. The *New York Times* reported on Wednesday, 29 October 1980, that eleven youths "who carried wooden crosses and shouted anti-Semitic slogans during a basketball game between Israeli and Italian teams were sentenced to up to three years and four months' imprisonment on charges of exalting genocide. . . . According to the testimony, the defendants, aged from their teens to 23, shouted slogans such as 'Jews to the ovens!' and 'Hitler taught us it's no crime to kill the Jews!' "

Sweden's Penal Code prohibits defamation of race. Chap. 16, sec. 8 (1972). In addition, the Instrument of Government (1G) forbids associations of a military nature that involve the persecution of a particular group. 1G 2:14. Another law prohibits unauthorized military uniforms and armbands. SFS 1947:164. See Kenneth Lassan, "Group Libel versus Free Speech: When Big Brother *Should* Butt In," 23 Duq. L. Rev. 77 (1984).

Germany has an extensive set of criminal laws forbidding racist speech, and particularly neo-Nazi speech. While the German Constitution guarantees freedom of expression in Art. 5, sec. 2 of the article qualifies the right by adding that the principle of free speech finds its limitations "in the rules of general laws, in the rules concerning the protection of the youth and in the rules protecting personal honor." The German courts have struggled to interpret the meaning of "general laws." The German Constitutional Court has ruled that "general laws" allows restrictions on the right of free speech provided they are not directed toward any specific opinion and, in addition, that they promote societal values that enjoy a higher rank than the principle of freedom of speech. For details, see the comments on Art. 5 in Theodor Maunz, et al., *Das Grundgesetz* (1982). Art. 18 of the German Consitution provides that anyone who abuses freedom of opinion "in order to combat the free democratic basic order shall forfeit these basic rights."

Under this scheme of broad constitutional provisions, there exist a variety of criminal laws prohibiting Nazi activities. It is a criminal offense to use the signs, labels, or uniforms of "unconstitutional," that is, Nazi, organizations. Nazi uniforms and standard forms of greeting (*Heil Hitler!*) are outlawed by sec. 86 and 86a StGB 85.

It is also a crime to write, print, or distribute writings that glorify acts of violence against human beings and incite hatred. Sec. 131 StGB 85. Anyone who disturbs the peace by inciting hatred against parts of the population, acts of violence, or malicious libel may be punished by imprisonment. Sec. 130 StGB 85.

Thus, a man who spray-painted slogans and swastikas on an unmarked police car was convicted under StGB Art. 130 (incitement of racial hatred) and StGB Art. 86a (use of Nazi slogans). OLG Koblenz, 31 Monatsschrift für Deutsches Recht [M.D.R.] 334 (1977). The man had written *Juda Verrecke* (Die, Jew) in two places on the car and had painted several swastikas.

67. See Neier, *Defending My Enemy*, 78–79. For an example of the kind of free speech arguments made in the general periodical literature about a dispute like *Skokie*, see principally two articles written by Professor Carl Cohen: "The Case Against Group Libel," *The Nation*, 24 June 1978, 757; "Skokie: The Extreme Test," *The Nation*, 15 April 1978, 422. See also "Cohen vs. Skokie: An Exchange," *The Nation*, 6 May 1978, 522.

## Chapter 2

1. See generally Fred S. Siebert, "The Libertarian Theory," in *Four Theories of the Press* (Urbana, Ill.: University of Illinois Press, 1956), 39.

2.

> On such fundamental issues—representation and consent, the nature of constitutions and of rights, the meaning of sovereignty—and in such basic ways, did the colonists probe and alter their inheritance of thought concerning liberty and its preservation. To conceive of legislative assemblies as mirrors of society and their voices as mechanically exact expressions of the people; to assume, and act upon the assumption, that human rights exist above the law and stand as the measure of the law's validity; to understand constitutions to be ideal designs of government, and fixed, limiting definitions of its permissible sphere of action; and to consider the possibility that absolute sovereignty in government need not be the monopoly of a single all-engrossing agency but *(imperium in imperio)* the shared possession of several agencies each limited by the boundaries of the others but all-powerful within its own—to think in these ways, as Americans were doing before Independence, was to reconceive the fundamentals of government and of society's relation to government.

Bernard Bailyn, *The Ideological Origins of the American Revolution* (Cambridge: Harvard University Press, 1967), 230.

3. This has been a long-standing basis for justifying the concept of freedom of speech:

> The democratic doctrine of freedom of speech and of the press, whether we regard it as a natural and inalienable right or not, rests upon certain assumptions. One of these is that men desire to know the truth and will be desposed to be guided by it. Another is that the sole method of arriving at the truth in the long run is by the free competition of opinion in the open market. Another is that, since men will invariably differ in their opinions, each man must be permitted to urge, freely and even strenuously, his own opinion, provided he accords to others the same right. And the final assumption is that from this mutual toleration and comparison of diverse opinions the one that seems most rational will emerge and be generally accepted.

Carl Becker, *Freedom and Responsibility in the American Way of Life* (New York: Knopf, 1945), 33.

4. A well-known and influential summary of the "values" served by free speech is the following by Professor Thomas Emerson:

> The system of freedom of expression in a democratic society rests upon four main premises. These may be stated, in capsule form, as follows:
>
> First, freedom of expression is essential as a means of assuring individual self-fulfillment. The proper end of man is the realization of his character and potentialities as a human being. For the achievement of this self-realization the mind must be free. Hence suppression of belief, opinion, or other expression is an affront to the dignity of man, a negation of man's essential nature. Moreover, man in his capacity as a member of society has a right to share in the common decisions that affect him. To cut off his search for truth, or his expression of it, is to elevate society and the state to a despotic command over him and to place him under the arbitrary control of others.
>
> Second, freedom of expression is an essential process for advancing knowledge and discovering truth. An individual who seeks knowledge and truth must hear all sides of the question, consider all alternatives, test his judgment by exposing it to opposition, and make full use of different minds. Discussion must be kept open no matter how certainly true an accepted opinion may seem to be; many of the most widely acknowledged truths have turned out to be erroneous. Conversely, the same principle applies no matter how false or pernicious the new opinion appears to be; for the unaccepted opinion may be true or partially true and, even if wholly false, its presentation and open discussion compel a rethinking and retesting of the accepted opinion. The reasons which make open discussion essential for an intelligent individual judgment likewise make it imperative for rational social judgment.
>
> Third, freedom of expression is essential to provide for participation in decision making by all members of society. This is particularly significant for political decisions. Once one accepts the premise of the Declaration of Independence—that governments "derive their just powers from the consent of the governed"—it follows that the governed must, in order to exercise their right of consent, have full freedom of expression both in forming individual judgments and in forming the common judgment. The principle also carries beyond the political realm. It embraces the right to participate in the building

of the whole culture, and includes freedom of expression in religion, literature, art, science, and all areas of human learning and knowledge.

Finally, freedom of expression is a method of achieving a more adaptable and hence a more stable community, of maintaining the precarious balance between healthy cleavage and necessary consensus. This follows because suppression of discussion makes a rational judgment impossible, substituting force for reason; because suppression promotes inflexibility and stultification, preventing society from adjusting to changing circumstances or developing new ideas; and because suppression conceals the real problems confronting a society, diverting public attention from the critical issues. At the same time the process of open discussion promotes greater cohesion in a society because people are more ready to accept decisions that go against them if they have a part in the decision-making process. Moreover, the state at all times retains adequate powers to promote unity and to suppress resort to force. Freedom of expression thus provides a framework in which the conflict necessary to the progress of a society can take place without destroying the society. It is an essential mechanism for maintaining the balance between stability and change.

Thomas Emerson, *The System of Freedom of Expression* (New York: Vintage Books, 1970), 6–9. See also Tribe, *American Constitutional Law*, (see chap. 1, n. 58), 576–79. That the general theoretical structure of the First Amendment has remained pretty much intact from Emerson's statement to the present day is shown by the summary of free speech theory contained in a recent treatise on the First Amendment: Melville Nimmer, *Freedom of Speech* (New York: Matthew Bender, 1984), chap. 1.

The variations on these themes have produced an abundant literature. For a major, and comprehensive, philosophical critique of prevailing theories of free speech, see Frederick Schauer, *Free Speech: A Philosophical Enquiry* (Cambridge: Cambridge University Press, 1982).

5. Zechariah Chafee, *Free Speech in the United States* (Cambridge: Harvard University Press, 1941; rpt. New York: Atheneum, 1969), 33 (all references are to later printing).

6. Id.

7. Alexander Meiklejohn, "Free Speech and Its Relation to Self-Government," in *Political Freedom: The Constitutional Powers of the People* (New York: Harper & Row, 1948; rpt. New York: Oxford University Press, 1965), 3 (all references are to later printing).

8. Id. at 24–27. Emphasis in original.

9. Id. at 26.

10. For Meiklejohn public speech is that used by people to plan together for the general welfare. Private speech has a more selfish motivation behind it and is not directed at solving public issues: "There are, then, in the theory of the Constitution, two radically different kinds of utterances. The constitutional status of a merchant advertising his wares, of a paid lobbyist fighting for the advantage of his client, is utterly different from that of a citizen who is planning for the general welfare." Id. at 37. The distinction will be examined in more detail in chap. 5.

11. Meiklejohn, "The First Amendment as an Absolute," 1961 Sup. Ct. Rev. 245, 263.

12. The idea that free speech is important for a democratic political community was not new with Meiklejohn. In the First Amendment case law, it can be found earlier—for example, in Brandeis's concurrence in *Whitney v. California*: "Those who won our independence believe that freedom to think as you will and to speak as you think are means indispensable to the discovery and spread of political truth; that without free speech and assembly discussion would be futile." 274 U.S. 357, 375 (1927) (Brandeis, J., concurring).

Still, it is fair to say that the post-Meiklejohn period is more suffused with the idiom of a relationship to democracy. Important First Amendment theorists who have identified themselves as adherents to the Meiklejohn perspective include Alexander Bickel, *The Morality of Consent* (New Haven, Conn.: Yale University Press, 1975), 62; and Bork, "Neutral Principles and Some First Amendment Problems," 47 Ind. L.J. 1, 26 (1971).

13. 376 U.S. 254 (1964).

14. 376 U.S. at 274, quoting James Madison, *Elliot's Debates on the Federal Constitution* (1876), vol. 4, 568.

15. "If neither factual error nor defamatory content suffices to remove the constitutional shield from criticism of official conduct, the combination of the two elements is no less inadequate. This is the lesson to be drawn from the great controversy over the Sedition Act of 1798, 1 Stat. 596, which first crystallized a national awareness of the central meaning of the First Amendment." 376 U.S. at 273.

16. 376 U.S. at 270.

17. Kalven, "The New York Times Case: A Note on 'The Central Meaning of the First Amendment,'" 1964 Sup. Ct. Rev. 191.

18. Id. at 214. Kalven also commented on the impact *Sullivan* was likely to have on the development of First Amendment doctrine:

> It is not easy to predict what the Court will see in the *Times* opinion as the years roll by. It may regard the opinion as covering simply one pocket of cases, those dealing with libel of public officials, and not destructive of the earlier notions that are inconsistent only with the larger reading of the Court's action. But the invitation to follow a dialectic progression from public official to government policy to public policy to matters in the public domain, like art, seems to me to be overwhelming. If the Court accepts the invitation, it will slowly work out for itself the theory of free speech that Alexander Meiklejohn has been offering us for some fifteen years now. Id. at 221.

19. Id. at 209.

20. The author of the majority opinion in the *Sullivan* case, Justice William Brennan, subsequently confirmed Kalven's interpretation of the *Sullivan* opinion as adopting a Meiklejohn perspective. See Brennan, "The Supreme Court and the Meiklejohn Interpretation of the First Amendment," 79 Harv. L. Rev. 1, 11–19 (1965).

21. See, e.g., Time, Inc., v. Hill, 385 U.S. 374, 387–88 (1967); Red Lion Broadcasting Co., Inc., v. F.C.C., 395 U.S. 367, 390 (1969); Rosenbloom v. Metromedia, Inc., v. 403 U.S. 29, 41 (1971); Columbia Broadcasting System, Inc., v. Democratic Natl. Comm., 412 U.S. 94, 102 (1973); Pittsburgh Press Co. v. Pittsburgh Comm. on Human Relations, 413 U.S. 376, 381–82 (1973); Gertz

v. Robert Welch, Inc., 418 U.S. 323, 340 (1974); Miami Herald Publishing Co. v. Tornillo, 418 U.S. 241, 257 (1974); Cox Broadcasting Corp. v. Cohn, 420 U.S. 469, 492 (1975); Virginia State Board of Pharmacy v. Virginia Citizens Consumer Council, Inc., 425 U.S. 748, 765 (1976).

22. Kalven, supra n. 17, at 205.

23. John Stuart Mill made the concession regarding the legitimacy of restricting people from taking a self-assumed status of slavery. J. S. Mill, *On Liberty*, ed. C. V. Shields (New York: Bobbs-Merrill, 1956), 125. For another discussion of the point developed here regarding the minimal condition of speech for a democratic society, see Ronald Dworkin, "Is the Press Losing the First Amendment?" *New York Review of Books*, 4 December 1980, 53–54, reprinted in Ronald Dworkin, *A Matter of Principle* (Cambridge: Harvard University Press, 1985) 381.

24. "We can never be sure that the opinion we are endeavoring to stifle is a false opinion; and if we were sure, stifling it would be evil still." Mill, *On Liberty*, 21.

25. Id.

26. The following is as close as Meiklejohn comes to identifying the advantages to the community in informational terms of allowing extremist speech:

> Now, in [a town meeting] method of political self-government, the point of ultimate interest is not the words of the speakers, but the minds of the hearers. The final aim of the meeting is the voting of wise decisions. The voters, therefore, must be made as wise as possible. The welfare of the community requires that those who decide issues shall understand them. They must know what they are voting about. And this, in turn, requires that so far as time allows, all facts and interests relevant to the problem shall be fully and fairly presented to the meeting. Meiklejohn, "Free Speech," in *Political Freedom*, 26.

27. Chafee, *Free Speech in the United States*, 33.

28. See, e.g., *McCormick On Evidence*, ed. Edward W. Cleary (St. Paul, Minn.: West, 1984), 544–48, discussing Rule 403 of the Federal and Revised Uniform Evidence Rules, which recognizes the power of the judge to exclude evidence when its probative value is "outweighed by the danger of unfair prejudice, confusion of the issues, or misleading the jury. . . ."

See also Richard O. Lempert and Stephen A. Saltzburg, *A Modern Approach to Evidence* (St. Paul, Minn.: West, 1978), 210–13: "[E]vidence of other crimes is likely to be very prejudicial. . . . The jurors are likely to mistake the degree to which the evidence is relevant. If the jurors are sure that one crime leads to another, they may not give other evidence in the case, particularly that evidence which tends to exonerate the defendant, the weight which it deserves."

29. John Milton, *Areopagitica*, ed. Richard C. Jebb (Cambridge, Eng.: Cambridge University Press, 1918), 58.

30. 250 U.S. at 630.

31. Id. at 629.

32. Id. at 628.

33. Gitlow v. New York, 268 U.S. 652, 673 (1925) (Holmes, J., dissenting).

34. "[P]ersons confronted with Cohen's jacket were in a quite different posture than, say, those subjected to the raucous emissions of sound trucks blaring outside

their residences. Those in the Los Angeles courthouse could effectively avoid further bombardment of their sensibilities simply by averting their eyes." 403 U.S. at 21.

35. See 578 F.2d at 1207. See also 69 Ill.2d at 618–19.

36. John Stuart Mill provided the classic statement of the position in his *On Liberty* (16):

> But there is a sphere of action in which society, as distinguished from the individual, has, if any, only an indirect interest: comprehending all that portion of a person's life and conduct which affects only himself or, if it also affects others, only with their free, voluntary, and undeceived consent and participation. When I say only himself, I mean directly and in the first instance; for whatever affects himself may affect others through himself; and the objection which may be grounded on this contingency will receive consideration in the sequel. This, then, is the appropriate region of human liberty. It comprises, first, the inward domain of consciousness, demanding liberty of conscience in the most comprehensive sense, liberty of thought and feeling, absolute freedom of opinion and sentiment on all subjects, practical or speculative, scientific, moral, or theological. The liberty of expressing and publishing opinions may seem to fall under a different principle, since it belongs to that part of the conduct of an individual which concerns other people, but, being almost of as much importance as the liberty of thought itself and resting in great part on the same reasons, is practically inseparable from it. Secondly, the principle requires liberty of tastes and pursuits, of framing the plan of our life to suit our own character, of doing as we like, subject to such consequences as may follow, without impediment from our fellow creatures, so long as what we do does not harm them, even though they should think our conduct foolish, perverse, or wrong. Thirdly, from this liberty of each individual follows the liberty, within the same limits, of combination among individuals; freedom to unite for any purpose not involving harm to others: the persons combining being supposed to be of full age and not forced or deceived.

Mill's position has generated a wide debate that addresses generally the problem of establishing the limits of law in regulating behavior, and, in particular, behavior the majority of society regards simply as immoral. Generally speaking, that literature considers the problem of what "harm" behavior may cause and what claims of harm are properly cognizable by a free society when enacting legislation. Acts such as adultery, drunkenness, and homosexuality are the staple of these discussions, but "liberty of thought and speech" are typically incorporated into it as well. For a nineteenth-century response to Mill, we have Stephen's *Liberty, Equality, Fraternity* (see chap. 1, n. 22). In this century the classic debate is repeated in H. L. A. Hart's *Law, Liberty, and Morality* (Stanford, Calif.: Stanford University Press, 1963) and Patrick Devlin's *The Enforcement of Morals* (London: Oxford University Press, 1965).

In my judgment, this literature generally pays insufficient attention to the potential psychological costs (both individual and social) of some speech acts, as developed in this chapter. This is undoubtedly because the literature tends to lump speech together with a variety of conduct and to deal with all of it in a rather abstract way (perhaps necessarily, given the breadth of the behavior considered). In chap. 4 I shall comment on the potential social value involved in separating speech out for special toleration and actually rely, in doing so, on the strength of the feelings generated by speech acts, as developed in this chapter, and in part on the lack of consensus on the positions advanced in the literature about the proper limits of social regulation of behavior.

37. See, e.g., Emerson, *The System of Freedom of Expression*, 8–9: "This marking off of the special status of expression is a crucial ingredient of the basic theory for several reasons. . . . Secondly, expression is normally conceived as doing less injury to other social goals than action. It generally has less immediate consequences, is less irremediable in its impact."

38. 274 U.S. at 377.

39. Meiklejohn, "Free Speech," in *Political Freedom*, 28.

40. 250 U.S. at 630.

41. Mill, *On Liberty*, 7.

42. It is a commonplace of the psychological literature that people sometimes denounce most strongly those ideas or actions which hold some attraction for them. For a development of this idea in the context of obscenity, see Gaylin, "The Prickly Problems of Pornography," 77 Yale L.J. 579 (1968). Gaylin notes that under the law, "Preeminently obscenity must be prurient; that is, it must be sexually exciting." Id. at 582. Gaylin goes on to say, "I think their [the Justices'] difficulty lies in the somewhat priggish assumption that something which is 'disgusting' cannot be 'exciting.' It is here that the psychiatrist would consider the jurist naive." Id. at 583.

43. It may be interesting at this point to listen to what one of the early architects of the modern free speech idea said in 1890 about the need to limit some speech. In what is often described as the most famous law review article ever written, Louis Brandeis (then in practice in Boston) wrote of the importance of developing a legal cause of action for invasions of privacy in order to give individuals the means to seek redress against those who published highly embarrassing facts about their private lives. After noting how the legal protection afforded individuals had changed over the centuries, from that of protecting people against physical attack to that of protecting them against less physical but nonetheless significant injuries to their sensibilities, Brandeis spoke of the injury—both individual and social—sustained by publication of gossip:

> The intensity and complexity of life, attendant upon advancing civilization, have rendered necessary some retreat from the world, and man, under the refining influence of culture, has become more sensitive to publicity, so that solitude and privacy have become more essential to the individual; but modern enterprise and invention have, through invasions upon his privacy, subjected him to mental pain and distress, far greater than could be inflicted by mere bodily injury. Nor is the harm wrought by such invasions confined to the suffering of those who may be made the subjects of journalistic or other enterprise. In this, as in other branches of commerce, the supply creates the demand. Each crop of unseemly gossip, thus harvested, becomes the seed of more, and, in direct proportion to its circulation, results in a lowering of social standards and of morality. Even gossip apparently harmless, when widely and persistently circulated, is potent for evil. It both belittles and perverts. It belittles by inverting the relative importance of things, thus dwarfing the thoughts and aspirations of a people. When personal gossip attains the dignity

> of print, and crowds the space available for matters of real interest to the community, what wonder that the ignorant and thoughtless mistake its relative importance. Easy of comprehension, appealing to that weak side of human nature which is never wholly cast down by the misfortunes and frailties of our neighbors, no one can be surprised that it usurps the place of interest in brains capable of other things. Triviality destroys at once robustness of thought and delicacy of feeling. No enthusiasm can flourish, no generous impulse can survive under its blighting influence.

Warren and Brandeis, "The Right to Privacy," 4 Har. L. Rev. 193, 196 (1890).

44. Isaiah Berlin, "Two Concepts of Liberty," in his *Four Essays on Liberty* (New York and London: Oxford University Press, 1969), 118, 155.

45. E.g., in Pruneyard Shopping Center v. Robins, 447 U.S. 74 (1980), the California Supreme Court had interpreted the state constitution to require the shopping center to permit some political activity on its premises. The center argued to the United States Supreme Court that this violated its First Amendment rights, because the state was in effect making it, through the enforced toleration, appear to endorse the political beliefs of others. While the Court rejected this position (for good reason, I believe), no one dismissed the argument as spurious or psychologically untenable as a general proposition. In fact, in an earlier case, Intl. Assoc. of Machinists v. Street, 367 U.S. 740 (1961), involving a union member who objected to the union's use of his dues to support political activities he disapproved of, the majority (which decided the case in favor of the union member but on statutory grounds) recognized the constitutional claim as being of "the utmost gravity," and Justice Black wrote in a separate opinion that he would have found a constitutional violation. The point, of course, is not that these cases are perfectly parallel to the situation of a society confronting within its midst speech behavior it regards as wrongful and dangerous; many potential differences are immediately apparent. In *Street* the union member was forced to give money to causes he disapproved of (but is there a significant difference between giving money and making available public streets?) and only a single individual was involved, not a large group (but should the identity of a group be treated as less important?). The general point is simply the importance of recognizing that the nature of the injury we are speaking of for the general community is not something with which we are totally unfamiliar.

46. It was on the basis of an interpretation of Meiklejohn as arguing for the political and sovereign right of democratic citizens to receive all information relevant to political issues that Professor Thomas Scanlon constructed his broader theory of freedom of speech as constituting the protection of "individual autonomy." See Scanlon, "A Theory of Freedom of Expression," 1 Phil. & Pub. Aff. 204 (1972). Basing his theory on what he called the "Millian Principle," Scanlon argued that certain "harms" could not "be taken as part of a justification for legal restrictions" on acts of expression:

> These harms are: (a) harms to certain individuals which consist in their coming to have false beliefs as a result of those acts of expression; (b) harmful consequences of acts performed as a result of those acts of expression, where the connection between the acts of expression and the subsequent harmful act consists merely in the fact that the acts of expression led the agents to

> believe (or increased their tendency to believe) these acts to be worth performing. Id. at 213.

He subsequently described his objectives in the following terms:

> I undertook to defend this principle by showing it to be a consequence of a particular idea about the limits of legitimate political authority: namely that the legitimate powers of government are limited to those that can be defended on grounds compatible with the autonomy of its citizens—compatible, that is, with the idea that each citizen is sovereign in deciding what to believe and in weighing reasons for action. This can be seen as a generalized version of Meiklejohn's idea of the political responsibility of democratic citizens.

Scanlon, "Freedom of Expression and Categories of Expression," 40 U. Pitt. L. Rev. 519, 531 (1979). On reflection, however, he later felt compelled to abandon the theory as incorrect. Of primary concern to him was the realization or recognition that the pursuit of being "autonomous," at least in the sense of acquiring all relevant information needed for independent decision making or self-government, was not an exclusive, or in some instances even a primary, end of human life: "Additional information is sometimes not worth the cost of getting it. The Millian Principle allows some of the costs of free expression to be weighed against its benefits, but holds that two important classes of costs must be ignored. Why should we be willing to bear unlimited costs to allow expression to flourish ...?" Id. at 533.

Professor Scanlon continues to believe that fundamental human interests are at stake both in the act of expression and in having access to expression. He argues in favor of a generally strong level of protection for certain "categories" of expression, especially "political" expression, on the ground that "where political issues are involved, governments are notoriously partisan and unreliable. Therefore, giving government the authority to make policy by balancing interests in such cases presents a serious threat to particularly important participant and audience interests." Id. at 544.

47. See, e.g., The President's Commission on Law Enforcement and Administration of Justice Task Force Report: Crime and Its Impact—an Assessment (1967). The task force found that reports of violent crime in addition to reports of rising crime rates led to "a sense of crisis in regard to the safety of both person and property.... A few cases of terrible offenses can terrorize an entire metropolis, and rising crime rates in once safe areas can arouse new fears and anxieties." Id. at 855.

48. See, e.g., Emile Durkheim, *The Division of Labor in Society*, trans. George Simpson (New York: Free Press, 1960), 102–3; Kai T. Erikson, *Wayward Puritans* (New York: Wiley, 1966).

49. See Richard Bessel, *Political Violence and the Rise of Nazism* (New Haven, Conn.: Yale University Press, 1984), 95: [Nazi violence] "greatly affected the tenor of political life during the last years of the Weimar Republic. Largely as a consequence of the growth of a movement which did not shrink from physical violence, such activity became an increasingly accepted element of German politics and the Nazis' opponents were forced to fight even more on the Nazis' terms."

50. 578 F.2d at 1205. "The Village's third argument is that it has a policy of fair housing, which the dissemination of racially defamatory material could undercut. We reject this argument without extended discussion."

51. Albert J. Smith, mayor of Skokie, stated in his deposition that "Skokie is very vulnerable to racial, ethnic and religious disharmony." Statement of Albert J. Smith, Collin v. Smith, No. 77 C. 2982 (N.D. Ill.). Smith explained the sources of the various underlying tensions:

> Many of the Jewish newcomers came to Skokie from all-Jewish neighborhoods where they had little daily contact with Christian neighbors. Many of the Christian newcomers came from all-Christian neighborhoods where they had little or no contact with Jewish neighbors. Despite this or perhaps even because of it, the two groups have managed to achieve mutual accommodation and respect. However, the sensitivities are great and the potential for disharmony is ever present.

In addition to the Jewish-Christian problem, there had been a recent influx of "visible minorities":

> Every school district within the Village contains children who constitute a visible minority. This includes Japanese, Chinese, East Indian, Pakistanian and Latino persons. These new minorities are welcomed to the Village and are in fact encouraged to come here by the Village policy of "Freedom of Opportunity to All." All of them come from families which share a history of discrimination and oppression by the Caucasian community and this circumstance makes them particularly vulnerable to the activities prohibited by the ordinance.

Finally, Smith discussed black-white tensions:

> Some 18 years ago, the first Black family to move into Skokie was met by riotous neighbors and required a major display of official force to safeguard their lives and property. Under enlightened leadership of Village officials, churches, synagogues and influential citizens, many Black families have moved into Skokie without incident. However, we are aware that our residents are not immune to the passions and prejudices which lie beneath the surface of many middle-class White people.

52. For an inquiry into the importance to communities of using public institutions to reflect and shape community values, see Joseph L. Sax, "The Case for Retention of the Public Land," in *Rethinking the Public Lands*, ed. S. Brubaker (Washington, D.C: Resources for the Future, Inc., 1984).

53. On 25 April, 1978 the *New York Times* carried a story headlined Chicago, City of First Amendment Battles, about the *Skokie* trial, among other things. On the same page was a short story, "Florida Vandals Deface Temple and Rabbi's Car." In Dade County vandals had painted swastikas and the letters H-I-T-L-E-R on a rabbi's car while it was parked at a synagogue. At another synagogue windows were smashed and swastikas painted.

Nine days earlier the *New York Times* reported that intruders had painted anti-Semitic slogans on the walls of the Jewish Defense League. *New York Times*, 16 April, 1978, 35, col. 6.

54. Bickel, The Morality of Consent, 71.

55. See, e.g., Riesman, "Democracy and Defamation: Control of Group Libel," 42 Colum. L. Rev. 727–31 (1942). Riesman noted "the systematic avalanche of falsehoods which are circulated concerning the various groups, classes and races which make up the countries of the Western world." Id. at 727. He continued: "Such purposeful attacks are nothing new, of course.... What is new, however, is the existence of a mobile public opinion as the controlling force in politics, and the systematic manipulation of that opinion by the use of calculated falsehood and vilification." Id. at 728.

56. "From the murder of the abolitionist Lovejoy in 1837 to the Cicero riots of 1951, Illinois has been the scene of exacerbated tension between the races, often flaring into violence and destruction. In many of these outbreaks, utterances of the character here in question, so the Illinois legislature could conclude, played a significant part." 343 U.S. at 259. "In the face of this history and its frequent obligato of extreme racial and religious propaganda, we would deny experience to say that the Illinois legislature was without reason in seeking ways to curb false or malicious defamation of racial and religious groups, made in public places and by means calculated to have a powerful emotional impact on those to whom it was presented." 343 U.S. at 261.

57. For a critique on the self-fulfillment rationale of free speech, on the ground that it fails to explain why self-fulfillment in the limited context of speech activities should be specially singled out for protection, see Schauer, *Free Speech: A Philosophical Enquiry*, 47–72.

## Chapter 3

1. See Emerson, *The System of Freedom of Expression* (see chap. 2, n. 4), 10:

> [T]he apparatus of government required for enforcement of limitations on expression, by its very nature, tends toward administrative extremes. Officials charged with the duties of suppression already have or tend to develop excessive zeal in the performance of their task. The accompanying techniques of enforcement—the investigations, surveillance, searches and seizures, secret informers, voluminous files on the suspect—all tend to exert a repressive influence on freedom of expression. In addition, the restrictive measures are readily subject to distortion and to use for ulterior purposes.

See also Zechariah Chafee, *Government and Mass Communications* (Hamden, Conn.: Archon Books, 1965), 477: "The First Amendment embodied a very strong tradition that the government should keep its hands off the press. Every new governmental activity in relation to the communication of news and ideas, however laudable its purpose, tends to undermine this tradition and render further activities easier."

2. The most important work on the importance of broadly structuring First Amendment doctrine in order to ensure a "check" on government abuse of its own authority is Blasi, "The Checking Value in First Amendment Theory," 1977 Am. B. Found. Research J. 521.

3. See Kalven, "The New York Times Case: A Note on 'the Central Meaning of the First Amendment,' " 1964 Sup. Ct. Rev. 191, 213: "It must be rec-

ognized, of course, that a reason implicit in the breadth of the protection afforded speech is due to the judicial recognition of its own incapacity to make nice discriminations."

4. 376 U.S. at 279 (1964).

5. New York Times Co. v. United States, 403 U.S. 713 (1971).

6. In his book *Free Speech in the United States* (see chap. 2, n. 5), Chafee discusses the lack of protection for speech resulting from prosecutions under the Espionage Act (63–64):

> No one reading the simple language of the Espionage Act of 1917 could have anticipated that it would be rapidly turned into a law under which opinions hostile to the war had practically no protection. Such a result was made possible ... by the tremendous wave of popular feeling against pacifists and pro-Germans during the war. This feeling was largely due to the hysterical fear of spies and other German propaganda. All of us on looking back to 1917 and 1918 are now sure that the emotions of ourselves and every one else were far from normal. During a war all thinking stops.

The clamor for prosecutions was so strong at the time that:

> [T]he Attorney General about a month before the end of the war issued a circular directing district attorneys to send no more cases to grand juries under the Espionage Act of 1918 without first submitting a statement of facts to the Attorney General and receiving by wire his opinion as to whether or not the facts constituted an offense under the Act. "This circular," says Mr. O'Brian, "is suggestive of the immense pressure brought to bear throughout the war upon the Department of Justice in all parts of the country for indiscriminate prosecution demanded in behalf of a policy of wholesale repression and restraint of public opinion." Id. at 69.

7. For example, David Caute, in his book *The Great Fear*, summarizes a 1954 survey of popular attitudes:

> An extensive survey conducted under the direction of Professor Samuel A. Stouffer, of Harvard, and published in 1954, showed 52 percent of a national cross section in favor of imprisoning all Communists (other polls yielded an even higher percentage). Eighty percent wanted to strip all Communists of their citizenship. A poll taken in 1952 showed 77 percent of respondents agreeing that Communists should be banned from the radio. But this massive intolerance was not focused on Communists alone: 45 percent would not allow Socialists to publish their own newspapers, and 42 percent wanted to deny to the press the right to criticize the "American form of government."

David Caute, *The Great Fear* (New York: Simon & Schuster, 1978), 215.

8. In a recent article, Professor Vince Blasi develops the idea of how First Amendment doctrine might be structured in order to protect valued expression in times of great intolerance, or "pathology." See Blasi, "The Pathological Perspective and the First Amendment," 85 Colum. L. Rev. 449 (1985).

9. See Bailyn, *The Ideological Origins of the American Revolution* (see chap. 2, n. 2), 272–301.

10. Alexis de Tocqueville devoted several chapters of his great work *Democracy in America* to the reasons for fearing majority rule in the United States. He expressed grave concerns about the majority's nearly unbounded power:

Hence the majority in the United States has immense actual power and a power of opinion which is almost as great. When once its mind is made up on any question, there are, so to say, no obstacles which can retard, much less halt, its progress and give it time to hear the wails of those it crushes as it passes.

The consequences of this state of affairs are fate-laden and dangerous for the future.

Alexis de Tocqueville, *Democracy in America*, vol. 1, trans. George Lawrence (Garden City, N.Y.: Doubleday, 1969), 248. Speaking of the "tyranny of the majority," Tocqueville noted that his "greatest complaint against democratic government as organized in the United States is not, as many Europeans make out, its weakness, but rather its irresistible strength. What I find most repulsive in America is not the extreme freedom reigning there but the shortage of guarantees against tyranny." Id. at 252. The power retained by the majority greatly exceeds that of any king, for "a king's power is physical only, controlling actions but not influencing desires, whereas the majority is invested with both physical and moral authority, which acts as much upon the will as upon behavior and at the same moment prevents both the act and the desire to do it." Id. at 254.

11. Mill, *On Liberty* (see chap. 2, n. 23), 6.
12. Id.
13. Id.
14. Walter Bagehot, "The Metaphysical Basis of Toleration," in *The Works and Life of Walter Bagehot*, vol. 6 (1915), 220.
15. According to Holmes, there would seldom be occasions "when you cared enough" to stop dissident speech. "But if for any reason you did care enough, you wouldn't care a damn for the suggestion that you were acting on a provisional hypothesis and might be wrong." Gunther, "Learned Hand and the Origins of Modern First Amendment Doctrine: Some Fragments of History," 27 Stan. L. Rev. 719, 757 (1975).
16. Id. at 744.
17. Meiklejohn, "Free Speech," in *Political Freedom* (see chap. 2, n. 7), 43.
18. For a summary of some of the literature on the origins of periods of great intolerance, see Nagel, "How Useful Is Judicial Review in Free Speech Cases?" 69 Cornell L. Rev. 302, 304–5 (1984).
19. Mill, *On Liberty*, 10.
20. Id. at 21–22.
21. Bagehot, *Works and Life*, 220.
22. "Socrates' Defense (Apology)" in *The Collected Dialogues of Plato*, vol. 5 (Princeton, N.J.: Princeton University Press, 1961).
23. Mark Twain, *Europe and Elsewhere: Corn Pone Opinions* (New York: Harper, 1923), 406.
24. Mill, *On Liberty*, 22.
25. Id. at 22–23.
26. Bagehot, *Works and Life*, 221.
27. Id.
28. Id. at 221–22.
29. See, e.g., Erik H. Erickson, *Toys and Reasons* (New York: Norton, 1977).

30. Bagehot, *Works and Life*, 223–24.
31. Id. at 224.
32. Id.
33. See Gunther, supra n. 15, at 757.
34. See, e.g., FCC v. Pacifica Foundation, 438 U.S. 726, 761 (1978) (Powell, J., concurring): "In my view, the result in this case does not turn on whether Carlin's monologue, viewed as a whole, or the words that constitute it, have more or less 'value' than a candidate's campaign speech. This is a judgment for each person to make, not one for judges to impose upon him."
35. Gertz v. Robert Welch, Inc., 418 U.S. 323, 339 (1974).
36. 578 F.2d at 1200.
37. In more recent years there has been some effort on the Court to relax this approach to free speech cases. See, e.g., FCC v. Pacifica Foundation, 438 U.S. 726, 742–48 (1978); Young v. American Mini Theatres, Inc., 427 U.S. 50, 70 (1976). In both decisions, Justice John Paul Stevens wrote opinions (once for a plurality and once for a majority) indicating that the scope of First Amendment protection could vary with the "value" of the expression at issue.
38. See Roth v. United States, 354 U.S. 476, 511–12 (1957) (Douglas, J., dissenting): "The standard of what offends 'the common consensus of the community' conflicts, in my judgment, with the command of the First Amendment that 'Congress shall make no law . . . abridging the freedom of speech, or of the press.' " Time, Inc., v. Hill, 385 U.S. 374, 400 (1967) (Black, J., concurring): "The freedoms guaranteed by [the First] Amendment are essential freedoms in a government like ours. That Amendment was deliberately written in language designed to put its freedoms beyond the reach of government to change while it remained unrepealed" (footnote omitted). Barenblatt v. United States, 360 U.S. 109, 140 (1959) (Black, J., dissenting): "The First Amendment says in no equivocal language that Congress shall pass no law abridging freedom of speech, press, assembly or petition." Dennis v. United States, 341 U.S. 494, 590 (1951) (Douglas, J., dissenting): "[T]he command of the First Amendment is so clear that we should not allow Congress to call a halt to free speech except in the extreme case of peril from the speech itself."

Justice Black also defended his "absolutist" approach in a law review article. Black, "The Bill of Rights," 35 N.Y.U. L. Rev. 865 (1960):

> Neither as offered nor as adopted is the language of this Amendment anything less than absolute. . . .
>
> The Framers were well aware that the individual rights they sought to protect might be easily nullified if subordinated to the general powers granted to Congress. One of the reasons for adoption of the Bill of Rights was to prevent just that. Id. at 874–75.

39. For example, in New York Times Co. v. Sullivan, 376 U.S. 254 (1964), Justice Brennan relied heavily on a broad consensus that the Sedition Act of 1798 was unconstitutional. He stated:

> If neither factual error nor defamatory content suffices to remove the constitutional shield from criticism of official conduct, the combination of the

> two elements is no less inadequate. This is the lesson to be drawn from the great controversy over the Sedition Act of 1798... which first crystallized a national awareness of the central meaning of the First Amendment. Id. at 273.

He continued: "Although the Sedition Act was never tested in this Court, the attack upon its validity has carried the day in the court of history" (footnote omitted).

Professor Kalven (see Kalven, supra n. 3, at 206–7) observed that opinions regarding the act's constitutionality were less than uniform:

> [U]ntil its disposition by the *Times* case, the status of the Sedition Act of 1798 remained an open question. It has been a term of infamy in American usage, but sober judgments about its constitutionality have been few indeed. Many distinguished commentators—Corwin, Hall, and Carroll, for example — regarded the Sedition Act as constitutional, and Story might also be numbered among them.

Kalven also noted that Congress included language in the Espionage Act that was very similar to that found in the Sedition Act and that the Court's opinions in such cases as Beauharnais v. Illinois, 343 U.S. 250 (1952), had made no mention of the unconstitutionality of the concept of seditious libel.

40. The ways in which the government tends to be portrayed in First Amendment jurisprudence is not without complications. The problem is nicely illustrated by the case law dealing with the constitutionality of public access rules for the mass media. In Miami Herald v. Tornillo, 418 U.S. 241 (1974), the Court struck down a Florida statute that required newspapers in the state to publish the replies of any political candidate criticized in their columns. The Court supported its holding by evoking the image of the government as censor: "The choice of material to go into a newspaper, and the decisions made as to limitations on the size of the paper, and content, and treatment of public issues and public officials—whether fair or unfair—constitutes the exercise of editorial control and judgment. It has yet to be demonstrated how governmental regulation of this crucial process can be exercised consistent with First Amendment guarantees of a free press as they have evolved to this time." Id. at 258. Nowhere mentioned in the Tornillo opinion was Red Lion Broadcasting Co. v. FCC, 395 U.S. 367 (1969), a major decision under the First Amendment, which upheld, in the context of the broadcast media, rules under the so-called fairness doctrine that were similar in content to the Florida statute. In Red Lion, however, the Court depicted broadcasters as the potential censors—"There is no sanctuary in the First Amendment," the Court said, "for unlimited private censorship operating in a medium not open to all" (Id. at 392)—and the government as the public institution that protects the public's interest in maintaining an open marketplace of ideas—"[T]he people as a whole retain their interest in free speech by radio and their collective right to have the medium function consistently with the ends and purposes of the First Amendment. It is the right of the viewers and listeners, not the right of the broadcasters, which is paramount" (Id. at 390).

I have discussed these cases, and the general area of which they are a part, at length in Bollinger, "Freedom of the Press and Public Access: Toward a

Theory of Partial Regulation of the Mass Media," 75 Mich. L. Rev. 1 (1976), and Bollinger, "On the Legal Relationship between Old and New Technologies of Communication," 26 German Yearbook of International Law, 269 (1983).

41. I have dealt with this issue previously. See Bollinger, "Free Speech and Intellectual Values", 92 Yale L. J. 438, 439–44 (1983); Bollinger, "The Press and the Public Interest: An Essay on the Relationship between Social Behavior and the Language of First Amendment Theory," 82 Mich. L. Rev. 1447 (1984).

42. The strongest advocate of this position was Chafee:

> [I]t is useless to define free speech by talk about rights. The agitator asserts his constitutional right to speak, the government asserts its constitutional right to wage war. The result is a deadlock. Each side takes the position of the man who was arrested for swinging his arms and hitting another in the nose, and asked the judge if he did not have a right to swing his arms in a free country. "Your right to swing your arms ends just where the other man's nose begins." To find the boundary line of any right, we must get behind rules of law to human facts. In our problem, we must regard the desires and needs of the individual human being who wants to speak and those of the great group of human beings among whom he speaks. That is, in technical language, there are individual interests and social interests, which must be balanced against each other, if they conflict, in order to determine which interest shall be sacrificed under the circumstances and which shall be protected and become the foundation of a legal right. It must never be forgotten that the balancing cannot be properly done unless all the interests involved are adequately ascertained, and the great evil of all this talk about rights is that each side is so busy denying the other's claim to rights that it entirely overlooks the human desires and needs behind that claim. *Free Speech in the United States*, 31–32 (footnote omitted).

43. 274 U.S. 357 (1927).

44. Id. at 377, 376.

45. Meiklejohn, "Free Speech," in *Political Freedom*, 28.

46. Id. at 48.

47. For the suggestion that the intolerance impulse may have a significantly greater impact on speech than on other conduct, thereby justifying the special protections afforded the former, see Schauer, *Free Speech: A Philosophical Enquiry* (see chap. 2, n. 4), 80–86.

48. The most commonly cited examples of judicial abandonment of constitutional principles are Dennis v. United States, 341 U.S. 494, 510 (1951); and Korematsu v. United States, 323 U.S. 214, 220 (1944) (in which the Court upheld the internment of civilian Japanese Americans).

49. See Neier, *Defending My Enemy* (see chap. 1, n. 2), 72, 73:

> [T]he ACLU succumbed to some of the pressures of 1940 and after a bitter internal dispute, adopted a resolution . . . barring from its governing personnel members of "political organizations supporting totalitarian dictatorship in any country." . . . [I]ts real target was the Communist party. Soon after the resolution was adopted, the ACLU board implemented it by expelling one of its number. . . .
>
> The lengths to which national ACLU officials went in the 1950s in protecting the ACLU against Communist influence only became publicly known during

the summer of 1977. At that time, the ACLU obtained from the FBI the files the Bureau had maintained on the ACLU from its beginnings in 1920. The FBI files... disclosed that the ACLU's Washington office director, Irving Ferman, regularly informed the FBI about persons active in state affiliates of the ACLU who Ferman thought were espousing left-wing causes.

50. 403 U.S. at 24.
51. Neier, *Defending My Enemy*, 3.
52. Id. at 4.
53. Id.
54. Id. at 4–5.
55. Id. at 5.

## Chapter 4

1. Restatement (Second) of Torts, Sec. 559 (1965): "A communication is defamatory if it tends so to harm the reputation of another as to lower him in the estimation of the community or to deter third persons from associating or dealing with him." Id. Sec. 652D, Publicity Given to Private Life. See also Gertz v. Robert Welch, Inc., 418 U.S. 323 (1974), in which a lawyer was awarded damages for an article that said he was a "Leninist" and a "Communist-fronter." (Note: While the Supreme Court has actively restructured the law of libel, through constitutional interpretation, it has yet to indicate whether the common law tort of invasion of privacy, by publication of embarrassing facts, will survive constitutional attack.)

2. See, e.g., Talley v. California, 362 U.S. 60 (1960); NAACP v. Alabama, 357 U.S. 449 (1958).

3. Mill, *On Liberty* (see chap. 2, n. 23), 38–39.

4. Besides Mill's *On Liberty*, see Stephen, *Liberty, Equality, Fraternity* (see chap. 1, n. 22); Hart, *Law, Liberty, and Morality* (see chap. 2, n. 37); Devlin, *The Enforcement of Morals* (see chap. 2, n. 37); H. L. A. Hart, "Social Solidarity and the Enforcement of Morality," in his *Essays in Jurisprudence and Philosophy*, (Oxford: Oxford University Press, 1983), 248.

5. It will be recalled that in Skokie the city attempted to prohibit certain military dress, as well as to regulate demonstrative and expressive behavior. The Illinois Appellate Court modified the lower court's injunction of the demonstration and prohibited only the wearing and displaying of swastikas. 51 Ill. App. 3d 279, 9 Ill. Dec. 90, 366 N.E.2d 347 (1977): "[T]he tens of thousands of Skokie's Jewish residents must feel gross revulsion for the swastika and would immediately respond to the personally abusive epithets slung their way in the form of the defendants' chosen symbol, the swastika. The epethets [*sic*] of racial and religious hatred are not protected speech...." [citation omitted] 366 N.E.2d at 357. In reversing that part of the appellate court's decision, the Illinois Supreme Court quoted extensively from *Cohen* and concluded:

> The display of the swastika, as offensive to the principles of a free nation as the memories it recalls may be, is symbolic political speech intended to convey to the public the beliefs of those who display it. It does not, in our opinion,

> fall within the definition of "fighting words," and that doctrine cannot be used here to overcome the heavy presumption against the constitutional validity of a prior restraint.
>
> Nor can we find that the swastika, while not representing fighting words, is nevertheless so offensive and peace-threatening to the public that its display can be enjoined. We do not doubt that the sight of this symbol is abhorrent to the Jewish citizens of Skokie, and that the survivors of the Nazi persecutions, tormented by their recollections, may have strong feelings regarding its display. Yet it is entirely clear that this factor does not justify enjoining defendants' speech. 69 Ill.2d at 615.

6. For a rich account of the interplay between intolerance of speech and of aliens, read Chafee's *Free Speech in the United States* (see chap. 2, n. 5).

7. There were numerous reports throughout the hostage crisis of violence, and threatened violence, against visiting Iranians. The Carter Administration issued calls for "restraint," and the Department of Justice announced its intentions to investigate incidents for possible civil rights violations. See, e.g., "Demonstrations by U.S. Students Are Revived by New Iranian Move," *N.Y. Times*, 19 November 1979, A14, col. 1; "Americans Assail Iranians Rallying in Washington," *N.Y. Times*, 10 November 1979, A6, col. 4; "Iranians in U.S. Fear Retaliation on Two Fronts," *N.Y. Times*, 22 November 1979, A18, col. 1; "Civiletti, Citing 'Rule of Law,' Urges Restraint Towards Iranians in U.S.," 3 December 1979, A12, col. 3; "Student, 15, Is Beaten Over Iranian Heritage," *N.Y. Times*, 6 December 1979, B2, col. 5; "Islamic Center, in Rural Indiana, Feels Shock Waves of Iranian Crisis," *N.Y. Times*, 10 December 1979, A4, col. 4.

8. L. Tolstoy, *Resurrection*, trans. Rosemary Edmunds (New York: Penguin Books, 1966), 374–75.

9. *The Ethics of Aristotle*, trans. J. A. K. Thompson (New York: Penquin Books, 1953), 73.

10. See O. W. Holmes, *The Common Law*, ed. Mark DeWolfe Howe (Boston: Little, Brown, 1963), 36: "The first requirement of a sound body of law is, that it should correspond with the actual feelings and demands of the community, whether right or wrong. If people would gratify the passion of revenge outside of the law, if the law did not help them, the law has no choice but to satisfy the craving itself, and thus avoid the greater evil of private retribution. At the same time, this passion is not one which we encourage, either as private individuals or as law-makers."

11. For the argument that the modern state must inevitably implement value choices and cannot maintain a position of pure neutrality, see Shiffrin, "Liberalism, Radicalism, and Legal Scholarship," 30 U.C.L.A. L. Rev. 1103 (1983).

12. Chafee was particularly sensitive to the fact that ultimately it could not be law, constitutional or otherwise, that could dictate whether the United States could properly claim to be a tolerant society; that really depended, he said, on the general attitude of the people:

> If a community does not respect liberty for unpopular ideas, it can easily drive such ideas underground by persistent discouragement and sneers, by social ostracism, by boycotts of newspapers and magazines, by refusal to rent halls, by objections to the use of municipal auditoriums and schoolhouses, by

> discharging teachers and professors and journalists, by mobs and threats of lynching. On the other hand, an atmosphere of open and unimpeded controversy may be made as fully a part of the life of a community as any other American tradition. The law plays only a small part in either suppression or freedom. In the long run the public gets just as much freedom of speech as it really wants.

Chafee, *Free Speech in the United States*, 563–64.

13. Montaigne, "Of Husbanding Your Will," in *The Complete Works of Montaigne*, trans. Donald M. Frame (Stanford, Calif.: Stanford University Press, 1943), 769.

14. "Even though the Skokie march never took place, Skokie remains the symbolic battleground. During the fifteen months between the time the Nazis first announced their intention to march in Skokie and the time they called off their march after their legal right to hold it was upheld, it was the subject of a great public debate. That debate continues. Almost every daily newspaper in the country published editorials about Skokie.... Skokie has been an inflammatory issue on hundreds of call-in radio shows around the country. It has been a leading topic of sermons in churches and synagogues. United States senators have stated their views in the *Congressional Record*. Skokie has provoked fierce debates in schools, offices, community centers, old-age homes, restaurants, living rooms, and wherever else people gather." Neier, *Defending My Enemy* (see chap. 1, n. 2), 8.

15. The Jewish Defense League has been the subject of considerable concern and embarrassment to other Jewish organizations. Neier made the following observations:

> An unacknowledged but important factor in the deliberations of the major Jewish organizations was the radical Jewish Defense League, a small group, founded in 1968 by Rabbi Meir Kahane, regarded with contempt by many Jewish leaders....
>
> Like the Nazis, the JDL knows how to attract attention out of all proportion to the number of its adherents. The JDL now claims a membership of 19,000, up sharply during the year following the first announcement of the Skokie march. The publicity it gets is a product of its militance, its effective use of symbols, and its choice of the times and places of its demonstrations....
>
> The Jewish Defense League's two-word slogan—"Never Again"—speaks volumes. It says that anti-Semitism will never again go unchallenged. Nazis will never again rise. The Holocaust will never again happen. Jews will never again go like sheep to the slaughter. The big Jewish organizations may regard JDL as a bunch of hoodlums, but every Jew is stirred deeply by the vow of never again. And every Jewish organization's stand on Skokie has been shaped by the sense of militancy and resistance that is epitomized in the JDL slogan.

Neier, *Defending My Enemy*, 32, 33.

16. 578 F.2d at 1210.

17. Neier observed:

> The resistance of Skokie's Jews to a proposed demonstration in their town by American Nazis was a kind of delayed response of anger about the past. When German Nazis overran their towns in Eastern Europe, most Jews had not resisted.... Resistance seemed like madness....

> Inevitably, the survivors have been affected by a controversy brought to the surface in the United States by the publication of Hannah Arendt's 1963 book, *Eichmann in Jerusalem*. It provoked both intense discussion of the need for resistance and recriminations about cooperation between Jewish community leaders in Europe and the Nazis ... [and it] made Jewish cooperation in their own extermination a subject of popular discussion in this country....
>
> There were fierce attacks on Arendt's book, many of them based on the view that she was blaming the victims of the Holocaust for their extermination. Her purpose, as she made plain in the book, was to demonstrate that hardly any of the voices that should have been raised in moral protest against Nazism were to be heard in Germany or the territories conquered by the Reich. Where political and religious leaders did speak out against the Nazis, notably in a country such as Denmark, most Jews were saved. Those Jews who died, and those Jews who cooperated with the Nazis in the belief that they were making things easier for other Jews, were victims of the silence of Europe's moral leadership as they were victims of the Nazis. Neier, *Defending My Enemy*, 28–30.

Neier quoted Arendt:

> "Jewish officials could bc trusted to compile the lists of persons and of their property, to secure money from the deportees to defray the expenses of their deportation and extermination, to keep track of vacated apartments, to supply police forces to help seize Jews and get them on trains, until, as a last gesture, they handed over the assets of the Jewish community in good order for final confiscation. They distributed the Yellow Star badges.... The well-known fact that the actual work of killing in the extermination centers was actually in the hands of Jewish commandos had been fairly and squarely established by witnesses for the prosecution—how they had worked in the gas chambers and the crematories, how they had pulled the gold teeth and cut the hair of the corpses, how they had dug the graves and, later, dug them up again to extinguish the traces of mass murder; how Jewish technicians had built gas chambers in Theresienstadt, where the Jewish 'autonomy' had been carried so far that even the hangman was a Jew." Neier, *Defending My Enemy*, 29.

18. There is a large body of literature on the general phenomenon known as survivor guilt. See, e.g., Kai T. Erickson, *Everything in Its Path: Destruction of A Community in the Buffalo Creek Flood* (New York: Simon & Schuster, 1976), 169–73; Robert Jay Lifton, *Death in Life: Survivors of Hiroshima* (New York: Random House, 1967), 35–37, 489–99; Martha Wolfenstein, *Disaster: A Pyschological Essay* (Glencoe, Ill.: Free Press, 1957), 216–19.

19. Neier quoted a newspaper reporter, who was a Skokie native, recalling an argument over dinner at which the reporter had declared, " 'To hell with the First Amendment! We cannot allow Nazism to happen again. Not without protest. If I saw a Nazi walking down Oakton Street, I would not trust myself to restrain from violence.' We debated some more. Opposing the Nazis' right to march, I heard myself like Bull Connor opposing Freedom Marchers. Soon I was on the other side." Neier, *Defending My Enemy*, 58–59.

20. Montaigne, "Of the Art of Discussion," in *The Complete Works*, 703–4.

21. Neier observed:

> [A] tiny Nazi movement serves the purposes of organized Jewry.... [T]he major Jewish organizations understand that by appearing in the guise that is

ugliest to non-Jewish Americans and wearing uniforms against which non-Jews fought a war, Nazi anti-Semitism preempts the field.... The Nazis deter the expression of anti-Semitism in forms that might be more palatable to the American public and, therefore, more threatening to the Jews. Other anti-Semites might impose restraints on themselves for fear of being bracketed with the almost universally hated Nazis. A strong Nazi movement would be a great danger to Jews in the United States; a weak Nazi movement with no potential for growth has its uses. Neier, *Defending My Enemy*, 34.

22. This thought recalls one of the concerns raised by Holmes in his dissent in *Abrams*:

> In this case sentences of twenty years imprisonment have been imposed for the publishing of two leaflets that I believe the defendants had as much right to publish as the Government has to publish the Constitution of the United States now vainly invoked by them. Even if I am technically wrong and enough can be squeezed from these poor and puny anonymities to turn the color of legal litmus paper; I will add, even if what I think the necessary intent were shown; the most nominal punishment seems to me all that possibly could be inflicted, unless the defendants are to be made to suffer not for what the indictment alleges but for the creed that they avow—a creed that I believe to be the creed of ignorance and immaturity when honestly held, as I see no reason to doubt that it was held here, but which, although made the subject of examination at the trial, no one has a right even to consider in dealing with the charges before the Court. 250 U.S. at 629–30.

23. A seminal article supporting judicial review in the free speech area is Chafee, "Freedom of Speech in War Time," 32 Harv. L. Rev. 932 (1919). See also Emerson, *The System of Freedom of Expression* (see chap. 2, n. 4), 11–14.

A modern and powerful critique of the benefits of judicial review in protecting valuable speech is Nagel, "How Useful Is Judicial Review in Free Speech Cases?" 69 Cornell L. Rev. 302 (1984): "At a minimum, the systemic utility of judicial review in free speech cases has been a matter characterized far too much by convenient assumptions and cheery faith.... [T]he Court's program, taken as a whole, has done great damage to the public understanding and appreciation of the principle of free speech by making it seem trivial, foreign, and unnecessarily costly.... [T]hese drawbacks are, for the most part, inherent in the judicial process and therefore can be avoided only by generally avoiding judicial review...." Id. at 340.

24. For a forceful articulation of this ethic in a free speech context, see Bork, "Neutral Principles and Some First Amendment Problems," 47 Ind. L.J. 1 (1971): "[T]he Court's power is legitimate only if it has, and can demonstrate in reasoned opinions that it has, a valid theory, derived from the Constitution, of the respective spheres of majority and minority freedom. If it does not have such a theory but merely imposes its own value choices, or worse, if it pretends to have a theory but actually follows its own predilections, the Court violates the postulates of the Madisonian model that alone justifies its power. It then necessarily abets the tyranny either of the majority or of the minority.... The Supreme Court regularly insists that its results, and most particularly its controversial results, do not spring from the mere will of the Justices in the majority but are supported, indeed compelled, by a proper understanding of the Constitution

of the United States. Value choices are attributed to the Founding Fathers, not to the Court." Id. at 3–4.

25. This observation about the institutional position of the judicial branch in the political system was noted by Robert Burt in "Constitutional Law and the Teaching of Parables," 93 Yale L.J. 455 (1984): "The Court is, as many have observed, 'the least dangerous branch,' without direct power to command 'the sword or the purse.' [Footnote omitted.] This does not mean that the Justices are without power. It does mean that, when their power is directly and adamantly challenged, they are dependent on others' acts of faith, of good faith toward them." Id. at 475.

26. Robert Cover, *Justice Accused: Antislavery and the Judicial Process* (New Haven: Yale University Press, 1975).

27. Whitney v. California, 274 U.S. 357, 376 (1927) (Brandeis, J., concurring).

28. The phrase is Wigmore's, which was quoted in chap. 1.

> Hence, the *moral right of the majority to enter upon the war imports the moral right to secure success by suppressing public agitation against the completion of the struggle.* If a company of soldiers in war-time on their way to the front were halted for rest in the public highway, and a disaffected citizen, going among them, were to begin thus to harangue: "Boys! this is a bad war! We ought not to be in it! And you ought not to be in it—" the state would have a moral right to step promptly up to that man and smite him on the mouth. So would any well-meaning citizen, for that matter.

Wigmore, "Abrams v. U.S.: Freedom of Speech and Freedom of Thuggery in War-Time and Peace-Time," 14 Ill. L. Rev. 539, 554 (1920) (emphasis in original).

## Chapter 5

1. Harry Kalven, *The Negro and the First Amendment* (Columbus: Ohio State University Press, 1965).

2. Id. at 4.

3. Id. at 5.

4. Id. at 6.

5. 376 U.S. 254 (1964).

6. Kalven, "The New York Times Case: A Note on 'the Central Meaning of the First Amendment,'" 1964 Sup. Ct. Rev. 191.

7. Id. at 208, 213–14.

8. Id. at 221; Meiklejohn, "Free Speech," in *Political Freedom* (see chap. 2, n. 7).

9. Brennan, "The Supreme Court and the Meiklejohn Interpretation of the First Amendment," 79 Harv. L. Rev. 1, 11–19 (1965).

10. Bork, "Neutral Principles and Some First Amendment Problems," 47 Ind. L.J. 1, 26 (1971).

11. Bickel, *The Morality of Consent* (see chap. 2, n. 12), 62.

12. As noted in chap. 2, Meiklejohn subsequently gave this apparently severe limitation on the scope of the First Amendment an expansive interpretation, saying that literature and art would be protected under his theory because such speech was also "relevant" to political decision making. Nevertheless, Meiklejohn did in theory keep all protected speech umbilically tied to the concept of self-government.

13. Meiklejohn, "Free Speech," in *Political Freedom*, 37. Recent cases have extended First Amendment protection to commercial speech, rejecting Meiklejohn's relegation of commercial speech to Fifth Amendment protections. See, e.g., Virginia State Bd. of Pharmacy v. Virginia Citizens Consumer Council, Inc., 425 U.S. 748 (1976); Bigelow v. Virginia, 421 U.S. 809 (1975).

14. Meiklejohn, "Free Speech," in *Political Freedom*, 24, 25.

15. Id. at 26.

16. Chafee himself observed that one of the two interests protected by the First Amendment, besides the "individual interest, the need of many men to express their opinions on matters vital to them if life is to be worth living," was the "social interest in the attainment of truth, so that the country may not only adopt the wisest course of action but carry it out in the wisest way." Chafee, *Free Speech in the United States* (see chap. 2, n. 5), 33.

Brandeis said it more explicitly in his concurrence in Whitney: "Those who won our independence believed that... freedom to think as you will and to speak as you think are means indispensable to the discovery and spread of political truth; that without free speech and assembly discussion would be futile." Whitney v. California, 274 U.S. 357, 375 (1927) (Brandeis, J., concurring). See also n. 18, infra.

17. The Constitution states in relevant part that "for any Speech or Debate in either House, [Senators and Representatives] shall not be questioned any other Place." U.S. Const., Art. I, sec. 6.

18. See Meiklejohn, "Free Speech," in *Political Freedom*, 36. It appears that precisely this argument was made at the beginning of the nineteenth century. Professor Leonard Levy describes a new argument for free speech, made by John Thomson in 1801:

> He noted that Article One, Section Six, of the Constitution provided that members of Congress "shall not be questioned," that is, held legally liable, for any speech they might make; their remarks were clothed with an immunity that gave them the right to say whatever they pleased in their legislative capacities. Thomson then reasoned that if freedom of discussion was necessary for them, it was equally necessary, indeed more so, for their sovereigns, the people whom they represented. The electorate must pass judgment on the proceedings of Congress and insure that the government operated for the benefit of the government [*sic*]. For the fulfillment of their electoral duties and their responsibility to protect themselves, the people could not be denied access to any viewpoint. The agents of the people were accordingly powerless to abridge the freedom of speech or press. The intention of the framers of the First Amendment, Thomson concluded, was to guarantee that the people possessed "the same right of free discussion" as their agents.

Leonard W. Levy, *The Legacy of Suppression: Freedom of Speech and Press in Early American History* (Cambridge: Harvard University Press, 1960; rpt. New York:

Harper & Row, 1963), 296. Citing J. Thomson, "An Enquiry, Concerning the Liberty, and Licentiousness of the Press, and the Uncontrollable Nature of the Human Mind" (1801).

19. Meiklejohn also drew textual sustenance for his thesis from the wording of the due process clause of the Fifth Amendment. (No person shall... be deprived of life, liberty, or property, without due process of law....) He reasoned as follows: (1) The language of the First Amendment with respect to freedom of speech is an absolute prohibition against official or governmental abridgment. (Congress shall make no law... abridging the freedom of speech....) (2) The Fifth Amendment permits the abridgment of liberty as long as it is accomplished in a manner consistent with due process. (3) The term *liberty* in the due process clause includes speech. (4) Since, therefore, the First Amendment forbids completely the abridgment of freedom of speech, while the Fifth Amendment due process clause permits the abridgment of speech, one can conclude that the framers intended to divide speech into two classes. Those classes are the "freedom of speech"—public speech of the "citizen who is planning for the general welfare"—and the "liberty of speech"—the private speech of, e.g., the "merchant advertising his wares, of a paid lobbyist fighting for the advantage of his client." Meiklejohn, "Free Speech," in *Political Freedom*, 36–37. (For a moment Meiklejohn considers whether the clause of the First Amendment providing for the right to petition for redress of grievances indicates an intent to protect a mere private interest, thereby reflecting a similar intent on the freedom of speech clause. But he happily concludes that his interpretation is not thwarted since "such a petition, whatever its motivation, raises definitely a question of public policy. It asserts an error in public decision.... They ask, therefore, for reconsideration. And in doing so, they are clearly within the field of public interest." Id. at 38).

These arguments similarly fail to lend a degree of textual inevitability to Meiklejohn's claim for tolerance. Assuming that the due process clause permits the regulation of some speech, it is not at all clear that this category of speech is what Meiklejohn defined as "private" speech. The association of the term *liberty* with *property* in the same clause seems a very weak basis on which to conclude, as Meiklejohn did, that the Fifth Amendment allows Congress to regulate speech concerned with "individual possessions." Id. at 37. The term *liberty* may just as reasonably be thought to have a broader, or different, compass than the term *property* (that is to say, they need not share similar characteristics). Moreover, even if the First Amendment is taken to protect "public" speech, by virtue of its reflection off of the supposed meaning of the Fifth Amendment, the question remains why "public" speech should encompass speech that advocates subversion of the democratic process.

20. Meiklejohn, "Free Speech," in *Political Freedom*, 4. Chafee noted the same disturbing trends in his review of Meiklejohn's book:

> This is a timely book. The country seems to be suffering again from an epidemic of hysteria such as it underwent during the "Red Menace" of 1919–1920. Even men who recognize "that the dangers from subversive organizations at the time of World War I were much exaggerated" are so apprehensive of the dangers from subversive organizations today that they are once more seeking to fight objectionable ideas with long prison sentences

and heavy pecuniary penalties. If the First Amendment as construed by the Supreme Court prevents such suppression, then they propose to amend the First Amendment in order, so they say, "to preserve the *whole* Constitution." Chafee, Book Review, 62 Harv. L. Rev. 891, 892 (1949) (footnote omitted).

21. Meiklejohn, "Free Speech," in *Political Freedom*, 43.

22. Id. at 27–28.

23. Id. at 57.

24. See id. at 77.

25. See Wigmore, "Abrams v. U.S.: Freedom of Speech and Freedom of Thuggery in War-Time and Peace-Time," 14 Ill. L. Rev. 539, 556–57 (1920).

26. See Bork, supra n. 10, at 31. Professor Bork disagreed with the sweep of protection that Meiklejohn and Kalven would extend under a self-government theory of free speech; he proposed limiting protection to speech that was explicitly "political" in character.

27. Meiklejohn, "Free Speech" in *Political Freedom*, 51.

28. Id. at 10.

29. Id. at 28.

30. Meiklejohn spoke repeatedly of the need for protection of radical speech, and whenever he did so, his argument shifted from the need for information to the possibility of fearlessness:

> If, then, on any occasion in the United States it is allowable to say that the Constitution is a good document, it is equally allowable, in that situation, to say that the Constitution is a bad document. If a public building may be used in which to say, in time of war, that the war is justified, then the same building may be used in which to say that it is not justified. . . . These conflicting views may be expressed, must be expressed, not because they are valid, but because they are relevant. If they are responsibly entertained by anyone, we, the voters, need to hear them. When a question of policy is "before the house," *free men choose to meet it not with their eyes shut, but with their eyes open. To be afraid of ideas, any idea, is to be unfit for self-government.* Id. at 27–28 (emphasis added).

Praising Brandeis's remarks in *Whitney*, Meiklejohn added, "We Americans are not afraid of ideas, of any idea, if only we can have a fair chance to think about it." Id. at 48. See also id. at 77: "We are saying that the citizens of the United States will be fit to govern themselves under their own institutions only if they have faced squarely and fearlessly everything that can be said in favor of these institutions, everything that can be said against them."

31. See, e.g., Scanlon, "A Theory of Freedom of Expression," 1 Phil. & Pub. Aff. 204 (1972). I discussed the source of appeal of this line of First Amendment theory in chap. 2, as well.

32. See Meiklejohn, "Free Speech," in *Political Freedom*, 21, 24–28, 37, 42.

33. Id. at 73–75.

34. Id. at 74–75.

35. 1 Seneca the Elder, "On Mercy," in *Moral Essays*, trans. J. Basore (Cambridge: Harvard University Press, 1970), 356, 371–73. Another defense of freedom of expression that fits into this general strain of argumentation, though it is virtually always lumped together with other dissimilar defenses, is that of

Union or to change its republican form, let them stand undisturbed as monuments of the safety with which error of opinion may be tolerated where reason is left free to combat it." Thomas Jefferson, First Inaugural Address, 4 March, 1801, reprinted in *The Complete Jefferson*, ed. S. Padover (New York: Duell, Sloan & Pearce, 1943), 384, 385. Toleration becomes proof—a "monument"—of one's confidence, security, and authority.

36. See Meiklejohn, "Free Speech," in *Political Freedom*, 48.

37. Id. at 46.

38. Id. at 45–46.

39. See id. at 61.

40. For a more extended analysis of Holmes's legal attitudes in general, see Rogat, "The Judge as Spectator," 31 U. Chi. L. Rev. 213 (1964). Professor Yosal Rogat portrays Holmes as "detached," "disengaged," and without concern for the way of life he believed in. Id. at 255. "To a remarkable degree," Rogat writes, "Holmes simply did not care." He would "bow to the way of the world," because his view was that power lay at the root of human affairs and power was not negotiable. Rogat presents this picture as a contrast to the common view of Holmes as "skeptical" and "humble." "He may have said that nothing was true once and for all, but on any particular occasion in making any particular decison he was convinced that he was right." Id. at 251. His acquiescence, his fatalism, his detachment, and his lack of caring are what led him to perform his judicial functions as he did, with such tolerance.

What I am saying here in one sense conflicts with Professor Rogat's description and in another sense does not. It seems possible for someone like Holmes to believe quite deeply in the truth of his beliefs and yet to assume a position of tolerance toward contrary beliefs while still "caring" about his own beliefs. It is certainly possible for a person to feel the intensity of belief and simultaneously think it is folly to feel that way, to recommend to oneself and to others that the preferable course is that which is contrary to one's nature. The problem may lie in the ambiguity of the idea of skepticism. There are people who appear to believe virtually nothing quite easily, and there are also people who believe most fervently and yet realize that those inclinations are without rational foundations and try, on that basis, to modulate them.

41. 250 U.S. at 630.

42. The quotation is from Ralph Waldo Emerson's essay "Experience." *Emerson's Essays* (New York: Harper & Row, 1926) 298.

43. Edmund Wilson, "Justice Oliver Wendell Holmes," in *Patriotic Gore* (New York: Farrar, Straus & Giroux, 1962), 777.

44. Id.

45. Id. at 762.

46. Id. at 764.

47. Gunther, "Learned Hand and the Origins of Modern First Amendment Doctrine: Some Fragments of History," 27 Stan. L. Rev. 719 (1975).

48. Id. at 757.

49. See Meiklejohn, "Free Speech," in *Political Freedom*, 73–74.

50. Bickel, *The Morality of Consent*, 3–30.

51. Meiklejohn, "Free Speech," in *Political Freedom*, 86–87.

51. Meiklejohn, "Free Speech," in *Political Freedom*, 86–87.
52. Id. at 87.
53. Id. at 28.
54. Of Holmes's proposed placement of faith in the marketplace, Professor Bickel wrote:

> This is the point at which one asks whether the best test of the idea of proletarian dictatorship, or segregation, or genocide is really the marketplace, whether our experience has not taught us that even such ideas can get themselves accepted there, and that a marketplace without rules of civil discourse is no marketplace of ideas, but a bullring. *The Morality of Consent*, 76–77.

55. "When a nation is at war, many things that might be said in time of peace are such a hindrance to its effort that their utterance will not be endured so long as men fight and that no Court could regard them as protected by any constitutional right." 249 U.S. at 52.
56. See Gunther, supra n. 47, at 719; Chafee, supra n. 20, at 900–1001; Wilson, *Patriotic Gore*, 775.

It is worth noting here that Learned Hand seems to have developed a similar turnabout in his lifetime on the matter of judicial enforcement of free speech protection for radical expression, just as Holmes did. The only difference is that his reversal of position went the opposite way from that of Holmes. Hand began as a fervent supporter of liberal free speech doctrine with his well-known decision in Masses Publishing Co. v. Patten, 244 Fed. 535 (S.D.N.Y.), rev'd, 246 Fed. 24 (2d Cir. 1917), from which the so-called incitement test was derived. He ended up striking a very conservative posture with his opinion in United States v. Dennis, 183 F.2d. 201 (2d Cir. 1950), aff'd, 341 U.S. 494 (1951)—a view of free speech he embraced personally in his 1958 Holmes Lectures; see Hand, "The Bill of Rights" (1958), in which (in Professor Gunther's characterization) he viewed "the First Amendment as one of a set of moral adjurations, not as a judicially enforceable norm." Gunther, supra n. 47, at 752. What is especially interesting is that "Hand and Holmes shared a common philosophical outlook," which included a disbelief "in absolutes or eternal truths" and a commitment to "skepticism." Id. at 732–33. To Hand, "Tolerance is the twin of Incredulity," (id. at 756) and "our chief enemies are Credulity and his brother Intolerance." Id. at 766.

57. Gunther, supra n. 47, at 757.
58. New York Times Co. v. Sullivan, 376 U.S. 254, 270 (1964).
59. Id. at 271, quoting Cantwell v. Connecticut, 310 U.S. 296, 310 (1940).
60. Id. at 272–73 (citation omitted).
61. Id. at 273, quoting Craig v. Harney, 331 U.S. 367, 376 (1947).
62. Cohen v. California, 403 U.S. 15 (1971).
63. Id. at 15.
64. Id. at 18.
65. Id. at 21.
66. Id.
67. Id. at 25.
68. Id.

69. Id.
70. Berlin, *Four Essays on Liberty* (see chap. 2, n. 44), 122–31.
71. Id. at 131.

## Chapter 6

1. See Terminiello v. Chicago, 337 U.S. 1, 37 (1949) (Jackson, J., dissenting).
2. Schenck v. United States, 249 U.S. 47, 52 (1919).
3. Abrams v. United States, 250 U.S. 616, 630 (1919) (Holmes, J., dissenting).
4. Whitney v. California, 274 U.S. 357, 376 (1927) (Brandeis, J., concurring).
5. Id. at 377–78.
6. Meiklejohn, "Free Speech," *in Political Freedom* (see chap. 2, n. 7), 49.
7. Id.
8. Masses Publishing Co. v. Patten, 244 Fed. 535 (S.D.N.Y.), rev'd, 246 Fed. 24 (2d Cir. 1917).
9. Id. at 540.
10. Gunther, "Learned Hand and the Origins of Modern First Amendment Doctrine: Some Fragments of History," 27 Stan. L. Rev. 719, 729 (1975).
11. Brandenburg v. Ohio, 395 U.S. 444, 447 (1969) (footnote omitted).
12. Chaplinsky v. New Hampshire, 315 U.S. 568, 572 (1942).
13. See, e.g., Kalven, "The New York Times Case: A Note on 'The Central Meaning of the First Amendment,'" 1964 Sup. Ct. Rev. 191, 217–18.
14. Gertz v. Robert Welch, Inc., 418 U.S. 323, 340 (1974).
15. Bickel, *The Morality of Consent* (see chap. 2, n. 12), 73.
16. See Roth v. United States, 354 U.S. 476, 484 (1957):

> All ideas having even the slightest redeeming social importance—unorthodox ideas, controversial ideas, even ideas hateful to the prevailing climate of opinion—have the full protection of the guarantees, unless excludable because they encroach upon the limited area of more important interests. But implicit in the history of the First Amendment is the rejection of obscenity as utterly without redeeming social importance.

In 1973 the Court announced the test currently used to determine whether speech is obscene and therefore unprotected:

> The basic guidelines for the trier of fact must be: (*a*) whether "the average person, applying contemporary community standards" would find that the work, taken as a whole, appeals to the prurient interest; (*b*) whether the work depicts or describes, in a patently offensive way, sexual conduct specifically defined by the applicable state law; and (*c*) whether the work, taken as a whole, lacks serious literary, artistic, political or scientific value. Miller v. California, 413 U.S. 15, 24 (1973).

17. 343 U.S. 250 (1952).
18. Tribe, *American Constitutional Law* (see chap. 1, n. 58), 618: "Recent Supreme Court decisions, however, have made clear that the 'fighting words' exception to first amendment protection is to be narrowly construed.... A 'fighting words' statute is unconstitutional on its face if it is not limited to words which

'have a direct tendency to cause acts of violence by the person to whom, individually, the remark is addressed'..." Tribe continues, "[G]overnment authorities may not suppress otherwise protected speech if imminent spectator violence can be satisfactorily prevented or curbed with reasonable crowd control techniques." Id. at 621. Advocates of the fighting-words doctrine continue to quote *Chaplinsky* for the proposition that this type of speech has little or no social value. See, e.g., Rosenfeld v. New Jersey, 408 U.S. 902, 904 (Powell, Burger, Blackmun, J.J., dissenting).

19. See, e.g., Ely, "Flag Desecration: A Case Study in the Roles of Categorization and Balancing in First Amendment Analysis," 88 Harv. L. Rev. 1482, 1490–91 (1975).

20. For Supreme Court decisions in the area, which do not reflect perfect consistency on this general issue, see, e.g., Chaplinsky v. New Hampshire, 315 U.S. 568 (1942); Terminiello v. Chicago, 337 U.S. 1 (1949); Feiner v. New York, 340 U.S. 315 (1951); Gregory v. Chicago, 394 U.S. 111 (1969); Edwards v. South Carolina, 372 U.S. 229 (1963); Cox v. Louisiana, 379 U.S. 536 (1965); Cohen v. California, 403 U.S. 15 (1971). The phrase "heckler's veto" was introduced by Professor Kalven. See Kalven, *The Negro and the First Amendment* (see chap. 5, n. 1), 140–45.

21. See, e.g., Bickel, *The Morality of Consent*, 73–74:

> [T]he question is, should there be a right to obtain obscene books and pictures in the market, or to foregather in public places—discreet, but accessible to all—with others who share a taste for the obscene? To grant this right is to affect the world about the rest of us, and to impinge on other privacies and other interests.... Perhaps each of us can, if he wishes, effectively avert the eye and stop the car. Still what is commonly read and seen and heard and done intrudes upon us all, wanted or not, for it constitutes our environment.

22. In American Booksellers Assoc. v. Hudnut, 598 F. Supp. 1316 (S.D. Ind. 1984), the federal district court held unconstitutional, under the First Amendment, an Indianapolis ordinance that prohibited "pornography," defined as "the graphic, sexually explicit subordination of women, whether in pictures or in words." A similar ordinance was passed by the city council of Minneapolis but was vetoed by the mayor. See generally, "The Proposed Minneapolis Pornography Ordinance: Pornography Regulation Versus Civil Rights or Pornography Regulation as Civil Rights?" 11 Wm. Mitchell L. Rev. 39 (1985).

23. See Gaylin, "The Prickly Problems of Pornography," 77 Yale L.J. 579, 582–86 (1968).

24. See Riesman, "Democracy and Defamation: Control of Group Libel," 42 Colum. L. Rev. 727, 730–31 (1942):

> Our thinking is still in terms of the "individual" and the "state," and our law of defamation, such as it is, is conceived of only as a protection against individual injury, as the law of assault and battery is a protection for individual life and limb. Hence defamatory attacks upon social groups are pretty much outside the scope of existing law, and the discovery of an adequate defense for groups must cope not only with many technical obstacles but with the customary refusal of American law to appreciate the role of groups in the social process.

25. See U.S. Const., Amend. 1: Congress shall make no law respecting an establishment of religion....

26. A recent example of this is the public response to the debate regarding abortion during the 1984 presidential campaign. See "Abortion Issue Threatens to Become Profoundly Decisive," *N.Y. Times*, 14 October, 1984, E3, col. 1.; "Mixing Personal Morality and Public Policy," *N.Y. Times*, 20 September, 1984, A31, col. 1.; "Church vs. State: Historical Concern," *N.Y. Times*, 15 September, 1984, 29, col. 1. See also, "Governor Finds Chicago Prelate an Ally in Debate with O'Connor," *N.Y. Times*, 6 October, 1984, 26, col. 1.; "Kennedy Chides Church Leaders on Role of State," *N.Y. Times*, 11 September, 1984, 1 col. 2; "Excerpts from Kennedy's Remarks on Religion," *N.Y. Times*, 11 September, 1984, 26, col. 1; Lewis, "The Question is Law," *N.Y. Times*, 11 September, 1984, 31, col. 6; "Ferraro Says Religion Won't Influence Policy," *N.Y. Times*, 13 September, 1984, B16, col. 3; "Catholic Theologians Have Mixed Reactions to Cuomo's Notre Dame Talk," *N.Y. Times*, 17 September, 1984, 1213, col. 1; "Episcopal Bishop Says Officials Must Put Law before Tenets," *N.Y. Times*, 17 September, 1984, 1213, col. 1.

27. Supreme Court opinions in the area of religion make frequent reference to the "divisiveness of religion in politics," though I do not think it is given the full significance it deserves nor are its implications for understanding the limits of open discussion appreciated. In Lemon v. Kurtzman, 403 U.S. 602, 622–23 (1971), Chief Justice Burger wrote:

> Ordinarily political debate and division, however vigorous or even partisan, are normal and healthy manifestations of our democratic system of government, but political division along religious lines was one of the principal evils against which the First Amendment was intended to protect.... It conflicts with our whole history and tradition to permit questions of the Religion Clauses to assume such importance in our legislatures and in our elections that they could divert attention from the myriad issues and problems that confront every level of government.

See also Walz v. Tax Commn., 397 U.S. 664, 695 (1970): "[G]overnment participation in certain programs, whose very nature is apt to entangle the state in details of administration and planning, may escalate to the point of inviting undue fragmentation." Board of Educ. v. Allen, 392 U.S. 236, 254 (1968) (Black, J., dissenting): "The First Amendment's prohibition against governmental establishment of religion was written on the assumption that state aid to religion and religious schools generates discord, disharmony, hatred, and strife among our people...." Engel v. Vitale, 370 U.S. 421, 429 (1962): "[The early Americans] knew the anguish, hardship and bitter strife that could come when zealous religious groups struggled with one another to obtain the government's stamp of approval from each King, Queen, or Protector that came to temporary power." Zorach v. Clauson, 343 U.S. 306, 318–19 (1952) (Black, J., dissenting): "It was precisely because Eighteenth Century Americans were a religious people divided into many fighting sects that we were given the constitutional mandate to keep Church and State completely separate."

28. See, "Impartial Coverage of Crisis Infuriating Some in Britain," *N.Y. Times*, 11 May, 1982, 6, col. 1; "Misled on Falklands, British Press Says," *N.Y.*

*Times*, 29 July, 1982, 3, col. 4; "Israel's Press and the War," *N.Y. Times*, 6 July, 1982, 1, col. 1.

29. Schenck v. United States, 249 U.S. 47, 52 (1919).

30. See Kalven, "The New York Times Case: A Note on 'the Central Meaning of the First Amendment,' " 1964 Sup. Ct. Rev. 191, 205.

31. For a critique of the position that judicial enforcment of free speech has promoted social tolerance of speech, see Nagel, "How Useful Is Judicial Review in Free Speech Cases?," 69 Cornell L. Rev. 302 (1984).

32. This is a basic problem with a classic exchange on the dispute over whether free speech should be treated as an "absolute" or as subject to "balancing." See Frantz, "The First Amendment in the Balance," 71 Yale L.J. 1424 (1962); Mendelson, "On the Meaning of the First Amendment: Absolutes in the Balance," 50 Calif. L. Rev. 821 (1962), and Frantz, "Is the First Amendment Law?—A Reply to Professor Mendelson," 51 Calif. L. Rev. 729 (1963).

33. Neier, *Defending My Enemy* (see chap. 1, n. 2), 59.

34. See, "Testimony on the Psychological Effects of Racial Slurs," by David Gutman, Ph.D., Defendants' Exhibit 13 at 7–8:

> If the victim or the group with which he identifies have been exposed to situations in which racial insults were integral parts of a larger scenario of persecution and violence, then his subsequent reaction, to the racial slur alone, will also be particularly violent, and psychologically unsettling. I am now referring to a well-known psychological principle, delineated by the noted developmental psychologist, Heinz Werner: each component of a situation implicated in strong emotional arousal can, by itself, and at a later date, stimulate the sentiments that were once generated by the total situation.... Under the compulsion of the *pars pro toto* principle, the logical boundary between insult and blow, between epithet and murder—breaks down; such nice distinctions become the luxury of those who can laugh at scars, because they never felt a wound.

35. West German law prohibits Nazi uniforms, banners, patches, and forms of greeting. Sec. 86a StGB 85. See also chap. 1, n. 66.

36. See, e.g., Rowan v. Post Office Dept., 397 U.S. 728 (1970).

37. Additionally, we cannot overlook the fact, because it is well illustrated by the episode involved here, that much linguistic expression serves a dual communicative function: it conveys not only ideas capable of relatively precise, detached explication, but otherwise inexpressible emotions as well. In fact, words are often chosen as much for their emotive as their cognitive force. We cannot sanction the view that the Constitution, while solicitous of the cognitive content of individual speech, has little or no regard for that emotive function which, practically speaking, may often be the more important element of the overall message sought to be communicated. Cohen v. California, 403 U.S. 15, 25–26 (1971).

38. See Gaylin, "The Prickly Problems of Pornography," 77 Yale L.J. 579, 589 (1968).

39. Ely, supra n. 19.

40. 391 U.S. 367 (1968). O'Brien and three friends burned their draft cards on the steps of the South Boston Courthouse in protest against the Vietnam

War. O'Brien's conviction for failure to have possession of his draft card was upheld by the Supreme Court. The Court applied the following test:

> [A] government regulation is sufficiently justified if it is within the constitutional power of the Government; if it furthers an important or substantial governmental interest; if the governmental interest is unrelated to the suppression of free expression; and if the incidental restriction on alleged First Amendment freedoms is no greater than is essential to the furtherance of that interest. Id. at 377.

41. Ely, supra n. 19, at 1497.
42. Id. at 1497–98.
43. Id. at 1498.
44. Id.
45. Id. at 1504.
46. See generally Ely, "Legislative and Administration Motivation in Constitutional Law," 79 Yale L.J. 1205 (1970).
47. Lovell v. Griffin, 303 U.S. 444 (1938) (ordinance prohibiting distribution of leaflets in street without prior permission is void); Hague v. CIO, 307 U.S. 496 (1939) (ordinance forbidding public assembly in streets or parks and distribution of printed matter without permit is void); Schneider v. Town of Irvington, 308 U.S. 147 (1939) (city may not prohibit leaflet distribution because of a concern over littering). It should be noted, however, that the Court has not yet held that there exists a First Amendment right to access. Instead, the Court has insisted that states be "neutral" in their licensing systems.

## Chapter 7

1. Kalven, *The Negro and the First Amendment* (see chap. 5, n. 1), 13.
2. Bork, "Neutral Principles and Some First Amendment Problems," 47 Ind. L.J. 1, 24 (1971).
3. *Elliot's Debates on the Federal Constitution* (1876), vol. 4, 571, quoted in New York Times Co. v. Sullivan, 376 U.S. 254, 271 (1964).
4. Cantwell v. Connecticut, 310 U.S. 296, 310 (1940), quoted in 376 U.S. 254 at 271; Mill, *On Liberty* (see chap. 2, n. 23), 65 quoted in 376 U.S. 254 at 272.
5. Supra chap. 3, n. 42.
6. For a defense of the "absolute" terminology in free speech discourse that emphasizes its important role in indicating the appropriate attitude to take towards free speech disputes, see, e.g., Charles L. Black, "Mr. Justice Black, the Supreme Court, and the Bill of Rights," 222 *Harper's Magazine* 63 (1961): "In formal logic, the 'balancing' and the 'absolute' positions can be rendered as identical. The issue must therefore be which of them most naturally, in common understanding, suggests the form which we would wish the judicial process in such cases to take. Stressing of the 'balancing' terminology tends both to create an endless series of successful objections, on grounds of policy, to the prevalence of the Bill of Rights freedoms, and to inhibit the Court from interfering with legislative judgments in this field. The 'absolutist' view, taken sensibly, would

tend to carve out large areas of personal freedom to be enjoyed without regard to transient legislative views or the pressing necessity of shutting people up, or making them worship alike, or jailing them after a short-cut trial. One can understand the appeal of the latter alternative to Mr. Justice Black and to others who are concerned that the Bill of Rights was meant to have vastly important effects not always agreeable to the majority of the moment; and who are also convinced that these effects should take place (as is lucidly clear on the face of the text) by means of comprehensive prohibitions—real, binding prohibitions—on the legislative branch." Id. at 68.

Justice Black articulated his views in a well-known 1960 address at the New York University School of Law, later published as Black, "The Bill of Rights," 35 N. Y. U. L. Rev. 865 (1960): "The great danger of the judiciary balancing process is that in times of emergency and stress it gives government the power to do what it thinks necessary to protect itself, regardless of the rights of individuals. If the need is great, the right of Government can always be said to outweigh the rights of the individual. If 'balancing' is accepted as the test, it would be hard for any conscientious judge to hold otherwise in times of dire need.... It is my belief that there *are* 'absolutes' in our Bill of Rights, and that they were put there on purpose by men who knew what words meant, and meant their prohibitions to be 'absolute.'... To my way of thinking, at least, the history and language of the Constitution and the Bill of Rights... make it plain that one of the primary purposes of the Constitution with its amendments was to withdraw from the Government all power to act in certain areas—whatever the scope of those areas may be.... Nothing that I have read in the Congressional debates on the Bill of Rights indicates that there was any belief that the First Amendment contained any qualifications." Id. at 867, 874–75, 878, 880.

7. See, e.g., Schauer, *Free Speech: A Philosophical Enquiry* (see chap. 2, n. 4); Nagel, "How Useful Is Judicial Review in Free Speech Cases?" 69 Cornell L. Rev. 302 (1984).

8. Wellington, "On Freedom of Expression," 88 Yale L.J. 1105, 1114 (1979).

9. Kalven, "Upon Rereading Mr. Justice Black On the First Amendment," 14 U.C.L.A. L. Rev. 428, 429–30 (1967): "To begin with, he passes a major test for a great judge on free speech issues. He displays the requisite passion. The requirement is not so much a question of arguing for the preferred position thesis; it is rather that the judge respond to the fact that this is not just another rule or principle of law. Mr. Justice Black has for thirty years always risen to the occasion when a free speech issue was at stake; he has always been vigilant and concerned."

10. Fiss, "Kalven's Way," 43 U. Chi. L. Rev. 1, 4–7 (1975).

11. The literature on this issue is extensive. For access to the debate, the reader might begin with the writings of Owen Fiss, who has argued that the judge's role is to implement public values. See Fiss, "The Supreme Court, 1978 Term—Forward: The Forms of Justice," 93 Harv. L. Rev. 1 (1979): "[T]he equal protection clause... is as specific as the free speech clause... but neither is very specific. They simply contain public values that must be given concrete meaning and harmonized with the general structure of the Constitution.... The absence of textual specificity does not make the values any less real, nor any less impor-

tant. The values embodied in such non-textually-specific prohibitions as the equal protection and due process clauses are central to our constitutional order. They give our society an identity and inner coherence—its distinctive public morality. ... [T]he task of the judge is to give meaning to constitutional values, and he does that by working with the constitutional text, history, and social ideals. He searches for what is true, right, or just.... Only once we reassert our belief in the existence of public values, that values such as equality, liberty, due process, no cruel and unusual punishment, security of the person, or free speech can have a true and important meaning, that must be articulated and implemented—yes, discovered—will the role of the courts in our political system become meaningful, or for that matter even intelligible." Id. at 11, 17 (footnotes omitted). See also Fiss, "Objectivity and Interpretation," 34 Stan. L. Rev. 739 (1982): "[T]he judge interprets a prescriptive text and in so doing gives meaning and expression to the values embodied in that text.... [L]egal interpretations are constrained by rules that derive their authority from an interpretive community that is itself held together by the commitment to the rule of law." Id. at 755, 762.

A different view is found in John Hart Ely, *Democracy and Distrust: A Theory of Judicial Review* (Cambridge: Harvard University Press, 1980), 43-72: "Experience suggests that in fact there will be a systematic bias in judicial choice of fundamental values, unsurprisingly in favor of the values of the upper-middle, professional class from which most lawyers and judges, and for that matter most moral philosophers are drawn. People understandably think that what is important to them is what is important, and people like us are no exception. Thus the list of values the Court and the commentators have tended to enshrine as fundamental is a list with which readers of this book will have little trouble identifying: expression, association, education, academic freedom, the privacy of the home, personal autonomy, even the right not to be locked in a stereotypically female sex role and supported by one's husband. But watch most fundamental-rights theorists start edging toward the door when someone mentions jobs, food, or housing: those are important, sure, but they aren't *fundamental*. ... The idea that society's 'widely shared values' should give content to the Constitution's open-ended provisions... turns out to be at the core of most 'fundamental values' positions.... There are two possible reasons one might look to consensus to give content to the Constitution's open-ended provisions. One *might* say one was seeking to protect the rights of the majority by ensuring that legislation truly reflect popular values. If that were the purpose, however, the legislative process would plainly be better suited to it than the judicial. This leaves the other possible reason for the reference, to protect the rights of individuals and minority groups against the actions of the majority... [but] it makes no sense to employ the value judgments of the majority as the vehicle for protecting minorities from the value judgments of the majority." Id. at 63, 68–69 (footnotes omitted).

See also Brest, "The Fundamental Rights Controversy: The Essential Contradictions of Normative Constitutional Scholarship," 90 Yale L.J. 1063 (1981); Brest, "Interpretation and Interest," Stan. L. Rev. 765 (1982); R. Dworkin, *Taking Rights Seriously* (Cambridge: Harvard University Press, 1977), 81–130; Grey, "Do We Have an Unwritten Constitution?" 27 Stan. L. Rev. 703 (1975); Sandalow,

"Judicial Protection of Minorities," 75 Mich. L. Rev. 1162 (1977); Sandalow, "Constitutional Interpretation," 79 Mich. L. Rev. 1033 (1981); Wellington, "Common Law Rules and Constitutional Double Standards: Some Notes on Adjudication," 83 Yale L.J. 221 (1973).

12. Mill, *On Liberty*, 24.

13. Bickel, *The Morality of Consent* (see chap. 2, n. 12), 62.

14. 250 U.S. at 628–29.

15. 403 U.S. at 25.

## Chapter 8

1. See generally the materials collected in William L. Carey and Melvin Aron Eisenberg, *Corporations* (Mineola, N.Y.: Foundation Press, 1980), 219–23.

2. Voltaire, *Philosophical Dictionary* (New York: Penguin Books, 1972), 390.

3. I have offered the suggestion before, in "The Sedition of Free Speech," 81 Mich. L. Rev. 867 (1983).

4. Letter 168, *Letters of John Adams Addressed to His Wife*, ed. Charles Francis Adams, vol. 2 (Boston: Freeman & Bolles, 1841).

5. Samuel Johnson, *The Rambler*, No. 25, ed. W. J. Bate and A. B. Strauss, vol. 3 of the Yale Edition of the Works of Samuel Johnson (New Haven, Conn.: Yale University Press, 1969), 136–37.

6. In an 1895 address to a graduating class of Harvard University, Holmes spoke of "The Soldier's Faith":

> I do not know what is true. I do not know the meaning of the universe. But in the midst of doubt, in the collapse of creeds, there is one thing I do not doubt, that no man who lives in the same world with most of us can doubt, and that is that the faith is true and adorable which leads a soldier to throw away his life in obedience to a blindly accepted duty, in a cause which he little understands, in a plan of campaign of which he has no notion, under tactics of which he does not see the use.

Reprinted in *The Mind and Faith of Justice Holmes*, ed. Max Lerner (Boston: Little, Brown, 1943), 18, 20.

# Index